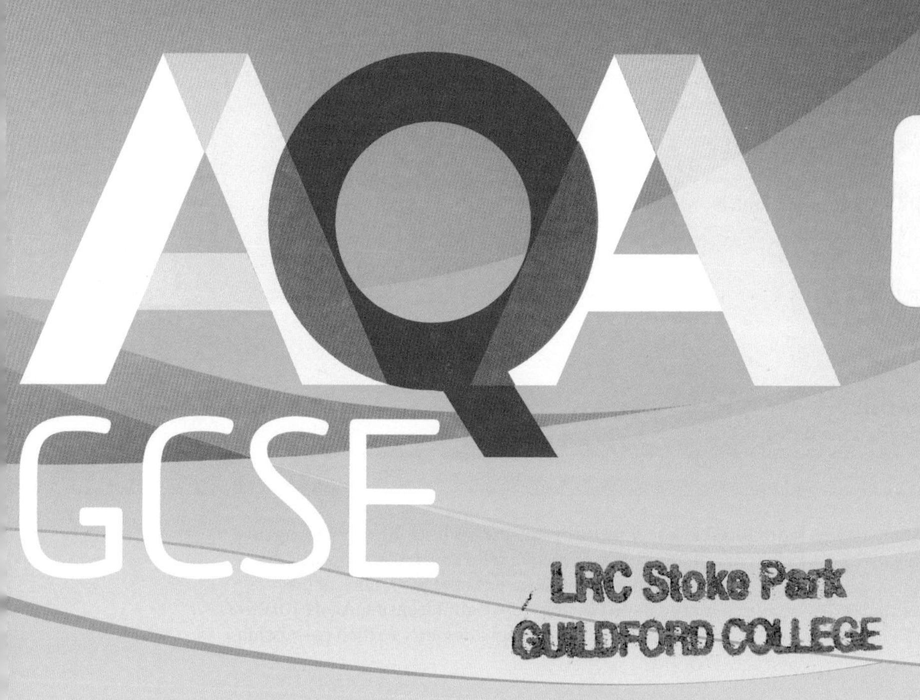

MATHEMATICS

Middle sets

PRACTICE BOOK

Series editor: **Glyn Payne**

Authors: **Greg Byrd, Lynn Byrd**

www.pearsonschools.co.uk

✓ Free online support
✓ Useful weblinks
✓ 24 hour online ordering

0845 630 22 22

Part of Pearson

Longman is an imprint of Pearson Education Limited, a company incorporated in England and Wales, having its registered office at Edinburgh Gate, Harlow, Essex, CM20 2JE. Registered company number: 872828

www.pearsonschoolsandfecolleges.co.uk

First published 2010
14 13 12 11 10
10 9 8 7 6 5 4 3 2 1

British Library Cataloguing in Publication Data
A catalogue record for this book is available from the British Library.
ISBN 978 1 408 23281 1

Edited by Anne Trevillion
Designed by Pearson Education Limited
Typeset by Tech-Set Ltd, Gateshead
Original illustrations © Pearson Education Ltd 2010
Illustrated by Tech-Set Ltd
Cover design by Wooden Ark
Cover photo © iStockphoto/Stuart Berman
Printed in the UK by Scotprint

Acknowledgements
Every effort has been made to contact copyright holders of material reproduced in this book. Any omissions will be rectified in subsequent printings if notice is given to the publishers.

Quick contents guide

UNIT 1 Statistics and Number

UNIT 2 Number and algebra Non-calculator

UNIT 3 Geometry and Algebra

2010 Specification changes: Assessment Objectives and functional maths

From 2010 the AQA GCSE Maths specifications have changed. For both Modular and Linear, the main features of this change are twofold.

Firstly the Assessment Objectives (AOs) have been revised so there is more focus on problem-solving. The new AO2 and AO3 questions will form about half of the questions in the exam. We provide lots of practice in this book, with AO2 and AO3 questions clearly labelled.

Secondly about 25% of the questions in the Higher exam, and about 35% of questions in the Foundation exam, will test functional maths. This means that they use maths in a real-life situation. Again we provide lots of clearly labelled practice for functional questions.

What does an AO2 question look like?

"**AO2** select and apply mathematical methods in a range of contexts."

An AO2 question will ask you to use a mathematical technique in an unfamiliar way.

3 Erik has a fair six-sided dice and a fair spinner numbered 1 to 4.
He rolls the dice and spins the spinner at the same time.
He multiplies the number on the dice and the number on the spinner to give the score.
What is the probability that he gets a score

a of 9   b of 12   c greater than 12?

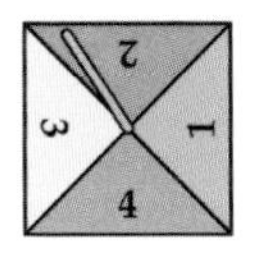

This just needs you to (a) read and understand the question and (b) use your knowledge of independant events to work out these probabilities. Simple!

What does an AO3 question look like?

"AO3 interpret and analyse problems and generate strategies to solve them."

AO3 questions give you less help. You might have to use a range of mathematical techniques, or solve a multi-step problem without any guidance.

7 Sandeep uses 1 cm^3 blocks to make three cubes. The first cube has a side length of 4 cm, the second cube has a side length of 5 cm and the third cube has a side length of 6 cm.
He breaks up the cubes and uses some of them to make the largest cube he can. How many blocks does he have left over?

C
AO3

Use your knowledge of cubes to work out, first, how many blocks he has in total...

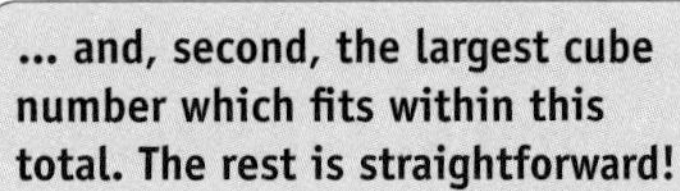

What does a functional question look like?

When you are answering functional questions you should plan your work. Always make sure that you explain how your answer relates to the question.

Read the question carefully.

7 Here is a recipe for salmon fishcakes.
Dai has 1.25 kg of salmon fillets and wants to use them all in some fishcakes.

a How much of each other ingredient does Dai need to make the fishcakes?

b How many people can Dai serve?

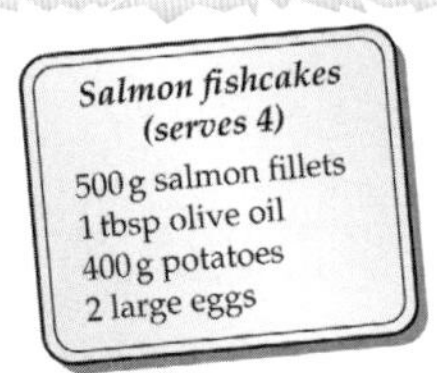

Think what maths you need and plan the order in which you'll work.

Follow your plan. Check your calculation. Job done!

How to use this book

This book has all the features you need to achieve the best possible grade in your AQA GCSE exams, **both Modular and Linear**. Throughout the book you'll find full coverage of Grades E-B, the new Assessment Objectives and Functional Maths.

At the end of the book you will find a **complete set of Practice Papers for Modular and a complete set for Linear**. We have included Higher Practice Papers for Modular Units 1 and 2, along with Foundation for Unit 3. We have included Higher Linear Practice Papers 1 and 2.

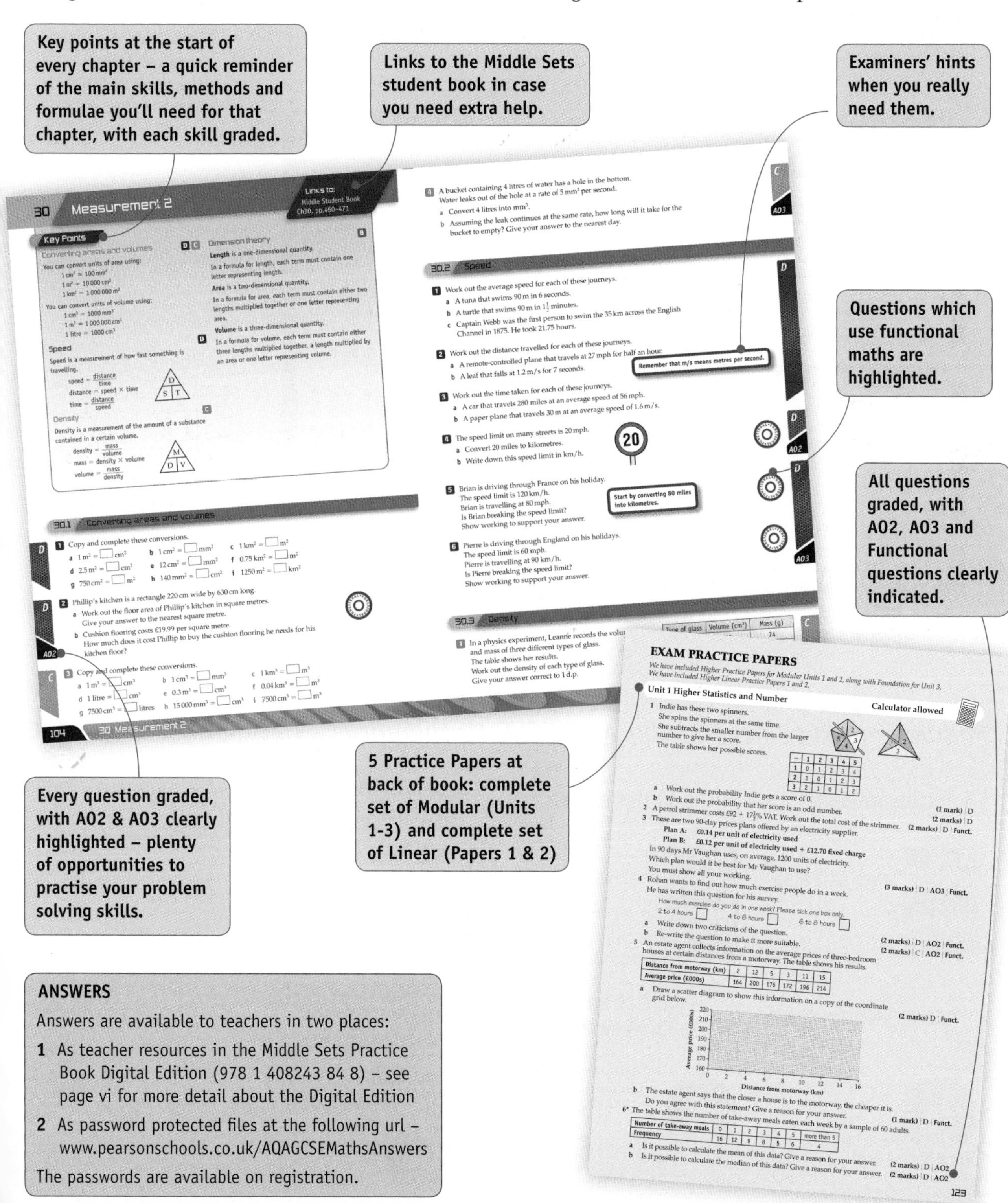

Key points at the start of every chapter – a quick reminder of the main skills, methods and formulae you'll need for that chapter, with each skill graded.
Links to the Middle Sets student book in case you need extra help.
Examiners' hints when you really need them.
Questions which use functional maths are highlighted.
All questions graded, with AO2, AO3 and Functional questions clearly indicated.
Every question graded, with AO2 & AO3 clearly highlighted – plenty of opportunities to practise your problem solving skills.
5 Practice Papers at back of book: complete set of Modular (Units 1-3) and complete set of Linear (Papers 1 & 2)
30 Measurement 2
Key Points
EXAM PRACTICE PAPERS

ANSWERS

Answers are available to teachers in two places:

1. As teacher resources in the Middle Sets Practice Book Digital Edition (978 1 408243 84 8) – see page vi for more detail about the Digital Edition
2. As password protected files at the following url – www.pearsonschools.co.uk/AQAGCSEMathsAnswers

The passwords are available on registration.

About the Digital Edition

We have produced a **Digital Edition** of the Middle Sets Practice Book (ISBN 978 1 408243 84 8) for display on an electronic whiteboard or via a VLE. The digital edition is available for purchase separately. It makes use of our unique **ActiveTeach** platform and will integrate with any other ActiveTeach products that you have purchased from the **AQA GCSE Mathematics 2010 series**.

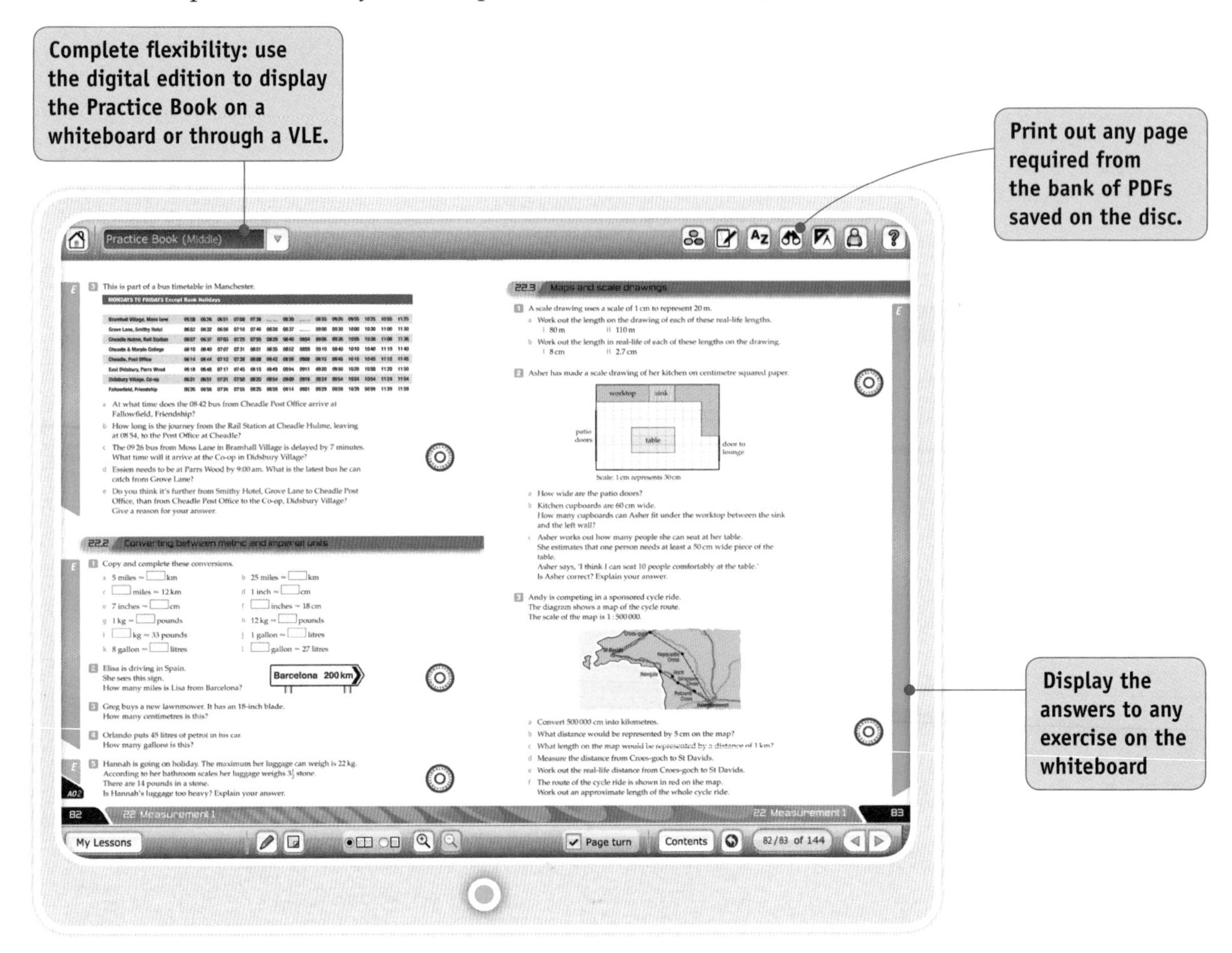

Print out any page required from the bank of PDFs saved on the disc.
Display the answers to any exercise on the whiteboard

Middle Sets Resources in the AQA GCSE Mathematics 2010 Series

STUDENT BOOK
E-B
9781408232828
PRACTICE BOOK
E-B
9781408232811
TEACHER GUIDE with EDITABLE CD-ROM
E-B
9781408232835
ACTIVETEACH CD-ROM
E-B
9781408232804
PRACTICE BOOK - Digital Edition
E-B
9781408243848
ASSESSMENT PACK with EDITABLE CD-ROM - Covering all sets
G-A*
9781408232842

1 Data collection

Links to:
Middle Student Book
Ch1, pp.1–18

Key Points

The data handling cycle D

A statistical investigation follows the data handling cycle:

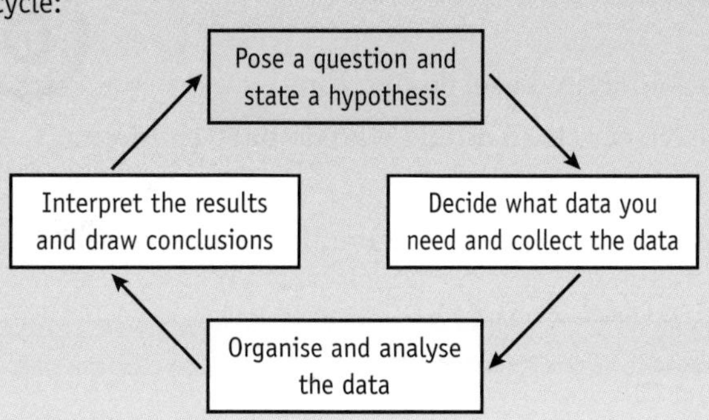

Stating a hypothesis D

A hypothesis is a statement that can be tested to answer a question.

A hypothesis must be written so that it is 'true' or 'false'.

Data sources D

Primary data is data you collect yourself.

Secondary data is data that has already been collected by someone else.

Types of data D

Qualitative data can only be described in words.

Quantitative data can be given numerical values and is either discrete or continuous.

Discrete data can only have certain values.

Continuous data can take any value in a range and can be measured.

Data collection E

A data collection table or frequency table has three columns: one for listing the items you are going to count, one for tally marks and one to record the frequency of each item.

Grouped frequency tables for discrete data D

Discrete data can be grouped into class intervals.

Class intervals should be equal sizes.

Grouped frequency tables for continuous data D

To group continuous data, the class intervals must use inequality symbols, $\leqslant$ and $<$.

$160\text{ cm} \leqslant h < 170\text{ cm}$ means a height from 160 cm up to **but not including** a height of 170 cm.

Recording data in a two-way table D

A two-way table helps you to present related data in a way that makes it easy to answer simple questions.

Questionnaires C

One way to collect primary data is to use a questionnaire.

A questionnaire is a form that people fill in.

Sampling techniques C

The total number of people you could ask to take part in a survey is called the population.

The smaller group of people you ask is called a sample.

A sample that is not representative of the population will be biased.

Random sampling allows every member of the population an equal chance of being selected.

1.1 The data handling cycle

D

1 'Young people are better at using calculators than old people.'
  a Give a reason why this is not a good hypothesis.
  b Rewrite the hypothesis to make it easier to test.

2 'Young people who don't get enough sleep can't learn properly.'
  a Give two reasons why this is not a good hypothesis.
  b Rewrite the hypothesis to make it easier to test.

3 'People under the age of 30 who live in my street enjoy eating curry.'
  a Give two reasons why this is a good hypothesis.
  b Rewrite the hypothesis, changing it to one that is not good.

1.2 Gathering information

D

1 Explain the difference between primary and secondary data.

2 For each hypothesis listed below, state:

- **i** whether you need primary or secondary data
- **ii** how you would find or collect the data
- **iii** how you would use the data.

FUNCTIONAL • FUNCTIONAL •

a In 2007, in the USA, more petrol engine vans were sold than diesel engine vans.

b Adults who don't have children drive more expensive cars than adults who do have children.

c Year 11 students eat a wider variety of vegetables than Year 7 students.

d Attendance at a newly built swimming pool is greater than at the old swimming pool it replaced.

1.3 Types of data

D

1 Use either number of pets or height as an example to explain the term 'discrete data'.

2 Use either number of pets or height as an example to explain the term 'continuous data'.

3 For each part below, write whether the data is quantitative or qualitative data.
If it is quantitative data, write whether it is discrete or continuous data.
Give an example of a typical item of data for each part.

a The weight of a car.

b The name of your favourite actor.

c The height of the back of a chair.

d The number of calculators in a maths classroom.

e The cost of a bunch of flowers.

f The colour of the pens in a pencil case.

g The length of wire needed to make a paper clip.

h Your birth date.

1.4 Data collection

E

1 Mignon rolled a fair six-sided dice. These are her results.

5	3	4	4	2	2	2	4	4	4	2	1
1	4	2	3	2	5	4	5	5	5	1	5
1	5	3	4	3	2	2	4	1	6	6	3

a Draw a frequency table to show this information.

b Which number did Mignon roll most often?

c How many times did Mignon roll the dice?

2 Some students were asked what was their favourite brand of sports-wear.
They had to choose from the following: Adidas (A), Umbro (U), Nike (N) and Reebok (R).

These are the results.

A	N	N	N	A	R	A	N	N	A	R	U
R	R	R	U	A	N	A	N	N	R	N	A

a Draw a frequency table to show this information.

b Which brand was chosen most often?

c How many students took part in the survey?

3 The teachers in a school were asked this question: 'How many holidays, including weekend breaks, do you usually go on each year?'
Their answers are given below.

4 3 6 1 0 4 5 5 6 2 1 7 5 2
4 5 4 3 3 2 6 0 5 3 3 1 4 3
2 6 7 3 6 3 6 4 4 4 1 4 5 6

a Draw a frequency table to show this information.

b What was the most common answer?

c Explain how you can use the frequency table to work out the total number of holidays taken by the teachers.

E

AO2

1.5 Grouped data

1 Here are the times taken by some wheelchair athletes to complete the 'Great North Run'.
The times are given to the nearest minute.

57 52 81 58 53 85 80 64 46 53 59 81
49 49 58 53 79 54 65 53 73 54 73 68

a Show the times in a grouped frequency table using the class intervals $40 \leqslant t < 45$, $45 \leqslant t < 50$, etc., where t represents the time in minutes.

b Which class interval contained the most times?

c How many athletes took less than 55 minutes?

d How many athlete's times are shown in the table?

D

2 An engineer writes reports for a motorcar magazine.
The engineer timed how long, to the nearest 0.1 second, it took various cars to go from 0 to 60 mph.
Here are his results.

4.8 4.7 6.3 8.5 8.4 14.2 9.7 7.5 11.8
6.2 4.2 5.5 12.6 13.9 9.3 10.4 12.0 13.6
5.0 4.6 9.4 8.8 5.5 7.3 9.4 8.3 8.4
5.7 10.9 14.6 7.8 7.5 6.0 13.7 14.8 8.3

Design a grouped frequency table to illustrate this data.
Choose suitable class intervals.

D

AO2

1.6 Two-way tables

1 This two-way table shows the age and gender of some young people at a youth club.

	Age (years)						
	11	12	13	14	15	16	Total
Boys	7	14	20	18	12	4	
Girls	12	22	17	8	5	2	
Total	19			26			

a Copy and complete the table.

b How many boys were older than 13?

c How many girls were younger than 13?

d How many young people were at the youth club?

D

D

2 This two-way table shows the results of the matches played by a darts team.

	Won	Drawn	Lost	Total
Home	6			
Away	9	2	8	
Total		5		32

a Copy the two-way table and fill in the missing numbers.

b How many away games were played by the darts team?

c How many games did the darts team lose?

d Five points are awarded for a win, two points for a draw and no points for a loss. How many points did the darts team have after playing all the games?

D

3 In a survey at a sports centre, 60 adults were asked what they had for breakfast. The responses were either cereal, cooked breakfast, toast or nothing.
11 women had cereal, 11 women had toast and 5 women did not eat anything.

There were 32 men in the survey. 17 of them had cereal, 5 had toast and 6 had a cooked breakfast.

a Design a two-way table to show this information.

b Complete the table showing the totals for men, women and each type of breakfast.

c How many of the adults surveyed had a cooked breakfast?

d How many of the adults surveyed had nothing for breakfast?

AO2

e What percentage of the adults surveyed had nothing for breakfast?

FUNCTIONAL • FUNCTIONAL •

1.7 Questionnaires

C

1 Arti wants to find out how much people spend in the supermarket each week.
She designs a questionnaire that includes these questions.

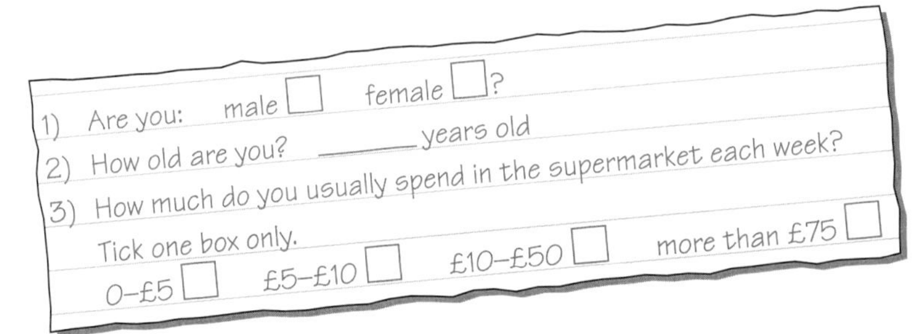
1) Are you: male ☐ female ☐?
2) How old are you? ______ years old
3) How much do you usually spend in the supermarket each week?
Tick one box only.
0–£5 ☐ £5–£10 ☐ £10–£50 ☐ more than £75 ☐

a Is question 1 a suitable or unsuitable question? Explain your answer.

b Is question 2 a suitable or unsuitable question? Explain your answer.

c Is question 3 a suitable or unsuitable question? Explain your answer.

C

2 Baz is carrying out a survey on the chocolate eating habits of students in his school.

a One of his questions is: 'On average, how many days a week do you have two or more snacks that contain chocolate?'
Design a response section for this question.

AO2

b Write a question that Baz can use to find out what is the favourite chocolate snack of the students.

3 Baz is carrying out another survey on the chocolate eating habits of students in his school. This time he wants to find out about differences in chocolate eating habits between

- each of the year groups in the school (Year 7 to Year 11)
- boys and girls.

Design a questionnaire and response sheet to gather data that includes

- the year group of the student
- the gender of the student
- the chocolate eating habits of the student.

C

4 Design a data collection sheet to show what students' favourite activities are on a Saturday morning.

AO2

1.8 Sampling

C

1 To find out if people prefer watching a film at the cinema or on DVD, a survey was carried out outside a 10-screen multiplex in the centre of town.
Is this likely to give a representative sample?
Give reasons for your answer.

2 To find out whether people think it is a good idea to ban smoking in public places, a survey was carried out in the canteen of a sports centre.
Why might this sample be unrepresentative?

3 A survey into students' favourite subjects was carried out by a geography teacher on her 'A' level geography students.
Do you think this will give a representative sample?
Explain your answer.

AO2

2 Fractions, decimals and percentages

Links to:
Middle Student Book
Ch2, pp.19–34

Key Points

Finding a fraction of an amount using a calculator E D

Most scientific calculators have a fraction key that looks like this [$a\frac{b}{c}$] or like this [$\frac{\blacksquare}{\square}$]. To find a fraction of an amount, enter the fraction, then multiply by the amount.

Writing one quantity as a fraction of another D

Make sure both quantities are in the same units, then write the first quantity over the second. Simplify the fraction when possible.

Calculating with fractions using a calculator E D

To carry out a calculation involving fractions, enter the fractions on your calculator using the fraction key.

Calculating the percentage of an amount E D

Divide the amount by 100 then multiply by the percentage you want to find.

Writing one quantity as a percentage of another D C

Write the first quantity as a fraction of the second, then multiply the fraction by 100 to convert it to a percentage.

Percentage increase and decrease D

To find the amount after a percentage change:

Method A

1. Work out the value of the increase (or decrease).
2. Add to (or subtract from) the original amount.

Method B

1. Add the percentage increase to 100% (or subtract the percentage decrease from 100%).
2. Convert this percentage to a decimal.
3. Multiply it by the original amount.

Retail prices index D C

An index number compares one number with another. It is a percentage of the base, which is usually 100.

2.1 Fraction of an amount

E

1 Use a calculator to work these out.

a $\frac{3}{4} \times £640$ b $3000 \times \frac{3}{8}$ c $\frac{3}{29}$ of 0.1305

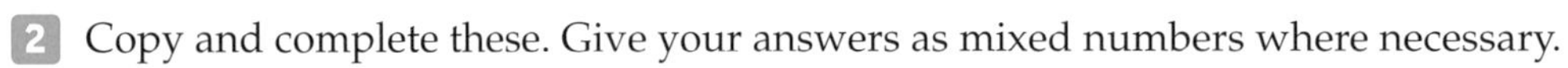

E AO2

2 Copy and complete these. Give your answers as mixed numbers where necessary.

a $\frac{3}{4} \times 1$ litre = $\square$ ml b $\frac{1}{3} \times 1$ year = $\square$ weeks
c $\frac{2}{7} \times 1$ week = $\square$ days d $\frac{3}{7} \times 1$ kg = $\square$ g

D

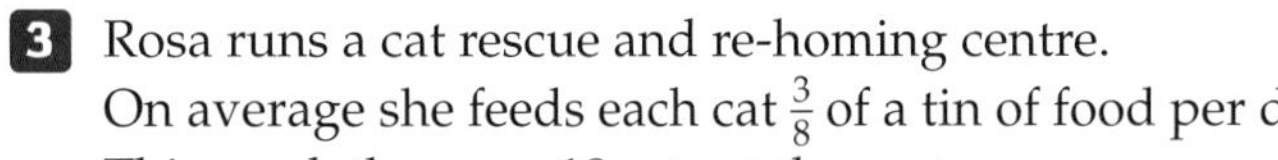

3 Rosa runs a cat rescue and re-homing centre.
On average she feeds each cat $\frac{3}{8}$ of a tin of food per day.
This week there are 18 cats at the centre.
How many tins of cat food does Rosa need for the week?

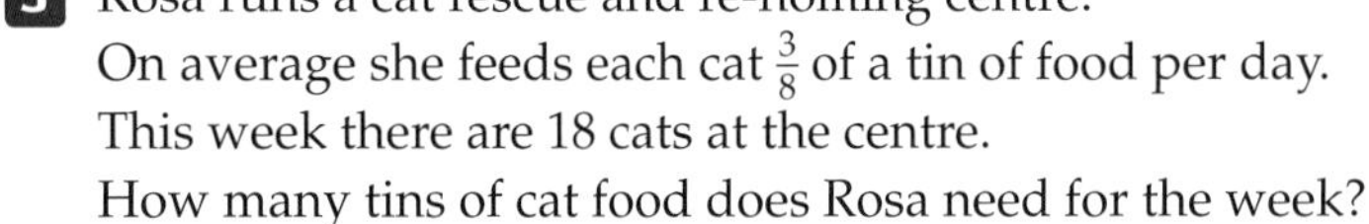

AO2

4 Greg makes a loaf of bread every 3 days. He uses $\frac{2}{5}$ of a bag of flour for each loaf.
How many bags of flour does he use in June?

2.2 One quantity as a fraction of another

D

1 Sandia gets 5 weeks' holiday a year.
What fraction of the year is Sandia on holiday?

2 Ali the anteater found a termite nest containing 1 000 000 termites.
He ate 35 000 of them. What fraction of the termites did Ali eat?

3 Yesterday was Brian's birthday.
He has spent £22 of his birthday money. He has £18 left.
What fraction of his birthday money has Brian spent?

D

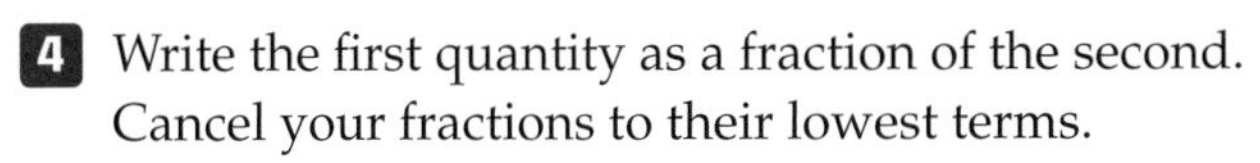

4 Write the first quantity as a fraction of the second.
Cancel your fractions to their lowest terms.

a £6, £36 **b** 6 m, 96 m **c** 6 kg, 9 kg

5 Write the first quantity as a fraction of the second.
Cancel your fractions to their lowest terms.

a 60 g, 9 kg **b** 6 m, 6 km **c** £5, £37.75

> **Make sure the units are the same before you start to simplify. Change both numbers in part c into pence first.**

D

6 Last year Mrs Sippy got 7 weeks' holiday and was also off sick for one week.
What fraction of last year was Mrs Sippy at work?

7 Colleen has a 3 m plank of wood to use for her technology project.
She cuts a piece 1.25 m long and paints it blue.
She cuts a piece 0.85 m long and paints it red.
She gives what is left to her friend Niall.
What fraction of the 3 m plank does Colleen

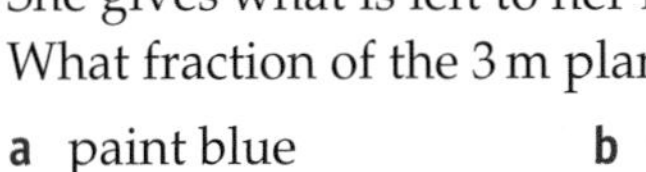

a paint blue **b** paint red **c** give to Niall?

AO2

2.3 Calculating with fractions

E

1 Use your calculator to work out the following. Give your answers as whole numbers.

a $\frac{1}{6} \times 96$ b $\frac{2}{7} \times 84$
c $\frac{3}{8} \times 192$ d $1080 \times \frac{4}{9}$

2 Use your calculator to work out the following.
Give your answers as fractions in their simplest form.

a $\frac{1}{6} \times 5$ b $\frac{2}{11} \times 5$
c $\frac{3}{18} \times 5$ d $5 \times \frac{4}{35}$

3 Use your calculator to work out the following.
Give your answers as mixed numbers.

a $\frac{1}{6} \times 23$ b $\frac{2}{7} \times 23$
c $\frac{3}{8} \times 23$ d $23 \times \frac{4}{9}$

4 Use your calculator to work out the following.
Give your answers as fractions in their simplest form.

a $\frac{1}{6} + \frac{2}{7}$ b $\frac{2}{11} - \frac{3}{18}$ c $\frac{2}{7} - \frac{3}{8} + \frac{4}{9}$

D

5 Use your calculator to work out the following.
Give your answers as mixed numbers.

a $3\frac{1}{6} - 1\frac{2}{7}$ **b** $7\frac{3}{11} + 4\frac{2}{18}$ **c** $4\frac{4}{7} - 1\frac{3}{8} + 3\frac{2}{9}$

6 A tea urn holds $24\frac{1}{2}$ litres of water.
A tea pot holds $1\frac{1}{6}$ litres of water.
How many teapots can be filled from one tea urn?

D

AO3

2.4 Percentages of amounts

E

1 Calculate the following. Write down your working.

a 16% of £400 b 5.5% of £400 c 0.15% of £400

d 210% of £400 e 1.01% of £4000 f 21% of £4

2 Last month Catrin bought a pair of jeans in a sale for £30.
Yesterday she saw the same pair of jeans. They cost 35% more than they did in the sale.
How much did Catrin save by buying her jeans in the sale?

3 Gina got 72% in her Spanish test. The test was out of 150.
How many marks did she score?

4 Sterling silver is made from 97.5% pure silver and 2.5% copper.
A sterling silver spoon weighs 28 g.
What mass of pure silver is in the spoon?

5 Before Christmas, Jorgan weighed 55 kg. After Christmas he weighed 12% more.
How much did he weigh after Christmas?

D AO3

6 Maha compares the prices of the same watch in two shops.
Which shop is cheaper and by how much?

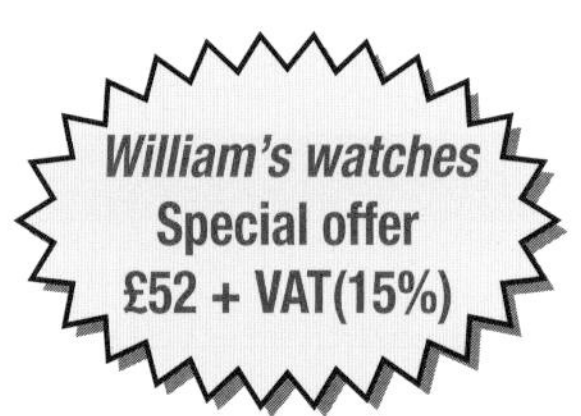

2.5 One quantity as a percentage of another

D

1 Ambika has 45 CDs. She lends nine of them to her friend.
What percentage of her CDs does Ambika lend to her friend?

2 Bob went shopping with £50. He spent £22.50.
What percentage of his money did Bob spend?

3 Chika went shopping with £50. She spent £32.50.
What percentage of her money did Chika not spend?

4 Yesterday Davin slept for 7 hours and had a nap for an hour in the afternoon.
What percentage of yesterday was Davin asleep?

C

5 A fat Jack Russell dog weighed 12.2 kg at the start of his diet. He now weighs 9.6 kg.
What percentage of his starting weight has he lost? Give your answer to 1 decimal place.

C AO2

6 A florist bought 125 white lilies from a market for 59p each.
The florist sold 74 of them for £1.75 each. The rest were not sold and were discarded.
Work out the florist's percentage profit.

2.6 Percentage increase and decrease

1 Bari used to earn £280 a week. He has had a 4% pay rise.
How much does Bari now earn?

2 In a sale the cost of a laptop is reduced by 30%.
Before the sale the laptop cost £499.
How much is the laptop in the sale?

3 An MP3 player costs £32 plus VAT at 15%.
What is the total cost of the MP3 player?

4 Clive bought a new car for £8750.
The value of the car went down by 15% in the first year and another 12% in the second year.
How much is the car worth at the end of the second year?

> Do not decrease the value by 27%.

5 James starts a new job earning £22 000 per year, increasing by 5% a year.
How much will James be earning after two years?

6 June weighed 72 kg.
Her weight increased by 5%.
June went on a diet and lost 5% of her body weight.
How much does June now weigh?

2.7 Index numbers

1 In December 2006 the total amount of money spent in shops using credit cards was £11.4 billion.
Using the year 2006 as the base year, the price index of December credit card spending is given for the years 2003 to 2007.

Year	2003	2004	2005	2006	2007
Index	92	100	101	100	103
Price				£11.4 billion	

Work out the December credit card spending from 2003 to 2007.

> An index of 92 means that the value has gone down by 8%. An index of 103 means that the value has gone up by 3%.

2 Compared to 2004 as a base, the index for DVD players this year is 38.

a Has the price of DVD players gone up or down?

b By what percentage has the price of the DVD players changed?

3 The retail prices index was introduced in January 1987. It was given a base number of 100.
20 years later, in January 2007, the index number was 204.8.
In January 1987 the 'standard weekly shopping basket' cost £38.50.
How much did the same 'standard weekly shopping basket' cost in January 2007?

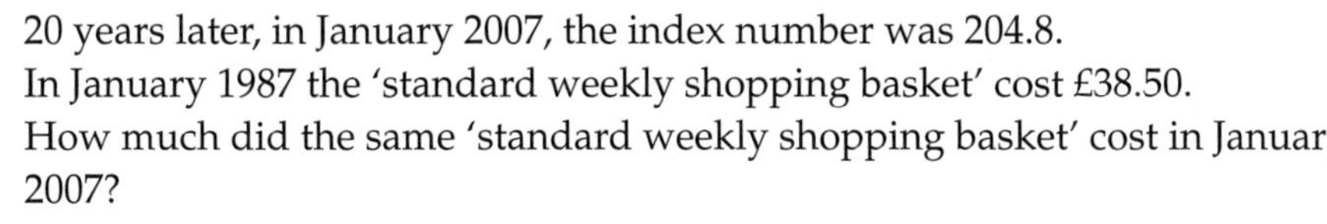

4 The graph shows the exchange rates for the euro and the pound from July 2008 to June 2009.

a What was the exchange rate in November 2008?

b Using July 2008 as the base of 100, work out the index for January 2009.

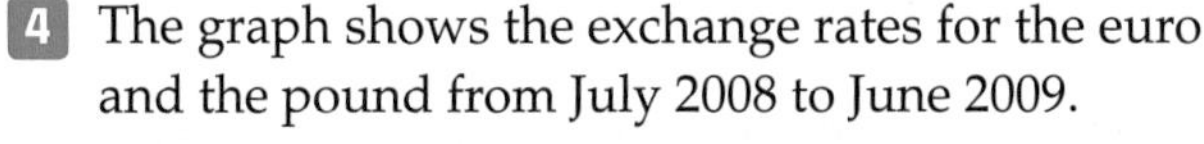

Exchange rate for euro to pound from July 2008 to June 2009

Number of euros to £1: 1.04, 1.08, 1.12, 1.16, 1.20, 1.24, 1.28

Months: J A S O N D J F M A M J

3 Interpreting and representing data

Links to:
Middle Student Book
Ch3, pp.35–52

Key Points

Drawing pie charts E

A pie chart is a circle that is split up into sectors. The angles in the sectors add up to 360°. The pie chart must be labelled and the angles accurately drawn.

Stem-and-leaf diagrams D

Stem-and-leaf diagrams are used to group and order data according to size, from smallest to largest. A key is needed to explain the numbers in the diagram.

Scatter diagrams and correlation D C

Scatter diagrams are used to compare two sets of data. They show if there is a connection or relationship, called a correlation, between the two quantities plotted.

On a scatter diagram a line of best fit is a straight line that passes through the data with an approximately equal number of points on either side, close to the line.

If a line of best fit can be drawn, then there is some form of linear correlation between the two sets of data.

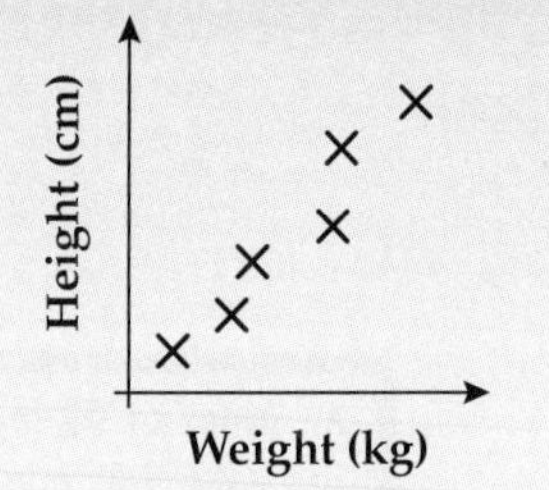

Positive correlation
As one quantity increases, the other quantity increases.

Price (£)
Age (years)

Negative correlation
As one quantity increases, the other quantity decreases.

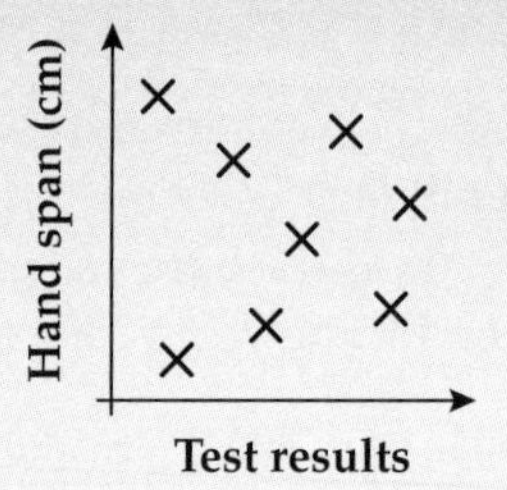

No (linear) correlation
Points are scattered randomly across the diagram with no (linear) correlation. This is also known as zero correlation.

Frequency diagrams for continuous data D

Continuous data can be represented by a frequency diagram. A frequency diagram is similar to a bar chart except it has no gaps between the bars.

Frequency polygons C

A frequency polygon shows patterns or trends in the data. When drawing a frequency polygon for grouped or continuous data, the mid-point of each class interval is plotted against the frequency.

3.1 Drawing pie charts

E

1 The frequency table shows how some students travel to school. Represent this data in a pie chart.

Transport	Frequency
car	45
bus	90
cycle	14
walk	31

2 Charles has collected data on the number of brothers that students in his class have. He has worked out the angles for his pie chart.
The frequency and angles are shown in the table.

Number of brothers	Frequency	Angle
0	9	60°
1	12	80°
2	3	20°
3	0	0°
4	3	20°

a Explain how you know that the angles Charles has calculated for his pie chart cannot be correct.

b Work out the correct angles for this data and draw a pie chart to represent the data.

E

AO2

3.2 Stem-and-leaf diagrams

D

1 A survey asked a group of newly qualified car drivers how many lessons they had taken before they passed their driving test. These are the results.

14 18 27 12 7 16 32 22 9 42

23 15 9 10 19 26 14 15 17 20

a Draw a stem-and-leaf diagram for this data.

b Use your diagram to work out how many drivers had taken fewer than 20 lessons.

2 In an experiment 20 girls were asked to draw a cube. The times they took, to the nearest tenth of a second, are shown below.

4.7 4.1 5.7 3.9 4.2 5.0 5.6 5.1 3.7 4.9

3.8 4.6 3.9 5.3 4.9 3.8 4.3 6.0 5.8 4.7

a Draw a stem-and-leaf diagram for this data.

b How many girls had a time of less than 5.5 seconds?

3.3 Scatter diagrams

D

1 The table shows the history and English exam marks of 15 students.

History	32	68	72	45	90	74	54	32	60	82	45	78	95	65	48
English	30	45	62	50	82	70	52	35	58	75	45	74	88	62	42

a Plot this information on a scatter diagram.

b Describe the relationship between students' history marks and English marks.

D

2 Dylis is investigating this hypothesis.
'The taller you are, the bigger your feet.'
She collects the following data from 10 people.

Height (m)	1.62	1.38	1.72	1.26	1.34	1.48	1.70	1.62	1.50	1.54
Shoe size	8	6	9	3	5	6	9½	7½	6	7

a Plot this information on a scatter diagram.

b Do you think Dylis's hypothesis is correct?
Give a reason for your answer.

AO2

D

3 Miss Trust is investigating this claim.

'The higher the percentage absence from her science lessons, the lower the mark in the science test.'

Miss Trust collects some data and draws a scatter diagram to show her results.

a Decide whether the claim 'The higher the percentage absence, the lower the mark in the science test' is correct.

b Students A and B do not fit the general trend. What can you say about

i student A

ii student B?

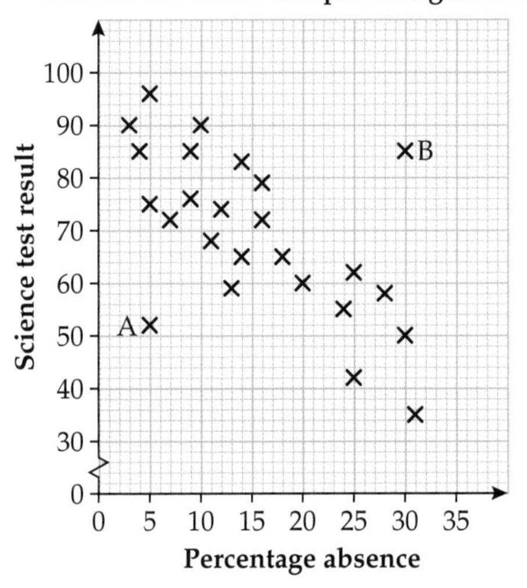

AO2

3.4 Lines of best fit and correlation

C

1 The table shows the maths and science exam marks of 15 students.

Maths	59	61	25	35	67	57	79	65	32	34	64	87	48	42	66
Science	55	57	23	34	34	45	64	61	33	30	58	78	44	41	62

a Draw a scatter diagram for this data.

b What type of correlation does the scatter diagram show?

c Draw a line of best fit on the scatter diagram.

d One student was ill when she took her science exam. Draw a circle around that piece of data on your scatter diagram.

e Another student was ill and missed her science exam. She scored 74% in her maths exam. Use your line of best fit to estimate the mark she would have had in her science exam.

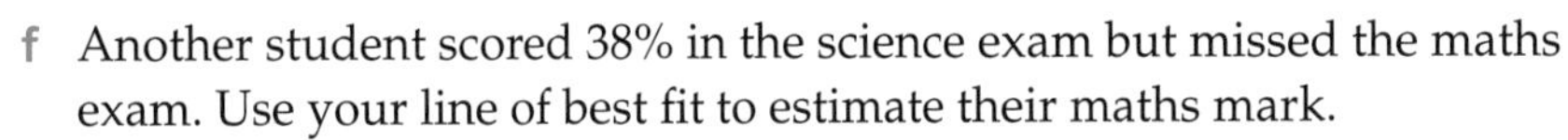

f Another student scored 38% in the science exam but missed the maths exam. Use your line of best fit to estimate their maths mark.

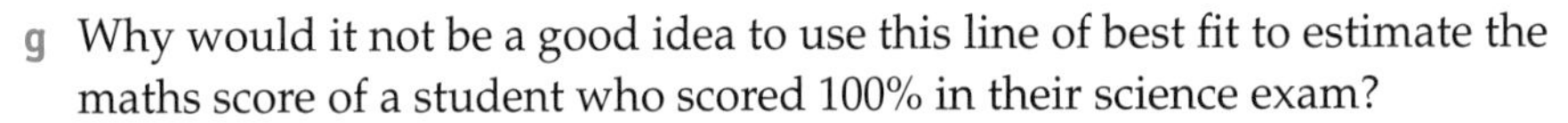

g Why would it not be a good idea to use this line of best fit to estimate the maths score of a student who scored 100% in their science exam?

AO2

3.5 Frequency diagrams for continuous data

D

1 The heights of 42 javelin throwers were recorded. The results are shown in the table.

Height, h (cm)	Frequency
$165 \leqslant h < 170$	2
$170 \leqslant h < 175$	6
$175 \leqslant h < 180$	9
$180 \leqslant h < 185$	14
$185 \leqslant h < 190$	8
$190 \leqslant h < 195$	3

Draw a frequency diagram to show this information.

2 A radio station carried out a survey on the ages of its listeners. The results are shown in the table.

D

Age, y (years)	Frequency
$10 \leqslant y < 20$	7
$20 \leqslant y < 30$	20
$30 \leqslant y < 40$	55
$40 \leqslant y < 50$	45
$50 \leqslant y < 60$	37
$60 \leqslant y < 70$	28

a Draw a frequency diagram to show this information.

b The radio producer is thinking of having a day when all the songs are hip-hop and rap. Do you think this is a good idea? Explain your answer.

AO2

3.6 Frequency polygons

1 The table shows the length of time, in seconds, that some smokers could hold their breath.

C

Time, t (seconds)	Frequency
$0 \leqslant t < 10$	1
$10 \leqslant t < 20$	9
$20 \leqslant t < 30$	16
$30 \leqslant t < 40$	7
$40 \leqslant t < 50$	5
$50 \leqslant t < 60$	1
$60 \leqslant t < 70$	1

Draw a frequency polygon for this data.

2 The two frequency polygons show the weight distributions of some long distance and some short distance runners.

C

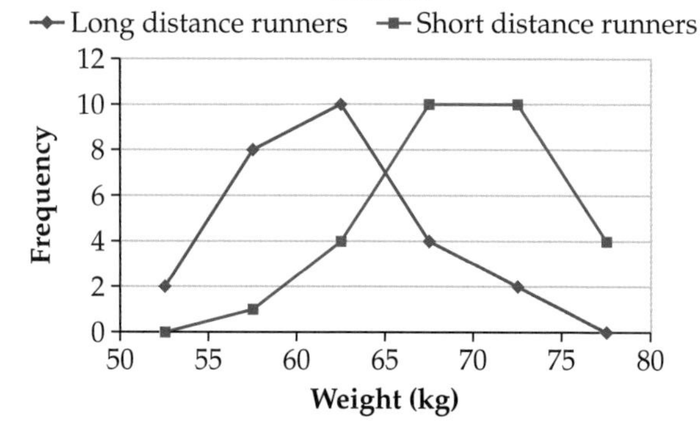

Compare the weight distributions of the two types of runners. Give a reason for your answers.

AO2

4 Range and averages

Links to:
Middle Student Book
Ch4, pp.53–68

Key Points

Range and averages E

The **mean** is the most commonly used average.

$$\text{Mean} = \frac{\text{sum of all the data values}}{\text{number of data values}}$$

The **median** is the middle value when the data is written in order. With an even number of pieces of data there are two middle values. The median is the value half way between them.

With n data values, you can work out the median using the formula

$$\text{Median} = \left(\frac{n+1}{2}\right)\text{th value}$$

The **mode**, or modal value, of a set of data is the number or item that occurs most often. Look for the highest frequency in a frequency table.

The **range** of a set of data is the difference between the largest value and the smallest value.

Range = largest value − smallest value

Calculating the mean from a frequency table D C

To calculate the **mean** from an ungrouped frequency table, you can use the formula

$$\text{Mean} = \frac{\text{sum of (data value} \times \text{frequency)}}{\text{total frequency}}$$

Range and averages from a grouped frequency table D C

When data is arranged in a grouped frequency table, you can't calculate the mean, mode, median and range exactly.

The class interval with the highest frequency is called the **modal class.**

You can estimate the **range** using the formula

Estimated range = (highest value of largest class interval) − (lowest value of smallest class interval)

With n data values, the **median** is the $\left(\frac{n+1}{2}\right)$th data value.

You can use this formula to work out which class interval contains the median.

Estimating the mean from a grouped frequency table C

You can estimate the **mean** by assuming that every data value lies exactly in the middle of a class interval. You need to work out the mid-point of each class interval.

$$\text{Mid-point of class interval} = \frac{\text{maximum value of class interval} + \text{minimum value of class interval}}{2}$$

$$\text{Estimate of mean} = \frac{\text{sum of (mid-point} \times \text{frequency)}}{\text{total frequency}}$$

4.1 Averages and range

E

1 Ellie grew eight sunflowers. The stem-and-leaf diagram shows the heights of the sunflowers in metres.

0	9
1	4 5 8 9
2	0 2 3

Key: 0 | 9 means 0.9 m

a Work out the range of these heights.

b Work out the mean height of the sunflowers.

c Work out the median height.

2 This table shows how much the Llewelyn family spent on holidays and weekend breaks last year.

Month	January	May	June	July	October	December
Cost (£)	238.50	521.75	135.00	1859.85	345.30	871.6

The family has a holiday savings account.
They pay the same amount into the account every month.
Use the mean to estimate how much they should pay into the account each month.

Notice that they do not go away every month, but they save the same amount every month.

3 The mean of five numbers is 4. Four of the numbers are 1, 2, 4 and 5.
Work out the fifth number.

E

4 The range of five numbers is 3.5. Four of the numbers are 2.7, 1.9, 2.1 and 3.9.
One number is missing.

a What is the smallest possible value for the missing number?

b What is the largest possible value for the missing number?

5 The mean of four numbers is 5.4. One of the numbers is 3.8.
Write down possible values for the other three numbers.

AO3

4.2 Calculating the range, mode and median from a frequency table

E

1 A hockey team recorded the number of goals they scored in each match in 2008.
The table shows the information.

Number of goals	0	1	2	3	4	5
Frequency	2	4	18	12	8	5

Work out

a the range of this data

b the modal number of goals scored

c the median.

2 A machine weighs apples as they pass through the cleaning station in a packing factory. The table shows the weight of some of the apples.

Weight (nearest 10 g)	Frequency
80	12
90	26
100	39
110	36
120	20

Work out

a the range of this data

b the modal weight of the apples

c the median.

E

3 This frequency table shows the number of letters in each word in one paragraph of a book.

Number of letters	1	2	3	4	5	6	7	8	more than 8
Frequency	8	20	17	5	17	9	10	6	3

Decide whether each of these statements is right or wrong.

If the statement is wrong, work out how to correct it.

a The mode of the data is 17.

b The range of the data is $8 - 1$ which is 7.

c The median of the data is 5.

AO3

4.3 Calculating the mean from an ungrouped frequency table

D

1 A hockey team recorded the number of goals they scored in each match in 2008. The table shows the information.

Number of goals	Frequency	Number of goals × frequency
0	2	$0 \times 2 = 0$
1	4	
2	18	
3	12	
4	8	
5	5	
Total		

a How many matches did the team play in 2008?

b Copy and complete the table to work out the total number of goals scored.

c Calculate the mean number of goals scored per match. Give your answer to one decimal place.

D C A02

2 Georgina works for a music company.
She is looking at data on music downloads from her company's website in December 2009.
This is the information she has.

Number of downloads in December 2009	Frequency		
	Age ranges:		
	10–19	20–29	30–39
1	37	32	53
2	20	34	27
3	6	39	19
4	15	47	12
5	32	33	2
6	12	6	0
7	3	0	1

a How many people aged 20–29 downloaded music in December 2009?

b Calculate the mean number of downloads per person in the 20–29 age range.
Give your answer to one decimal place.

c The company wants to target its advertising at the age range with the highest mean number of downloads. Which age range should it target?

4.4 Finding the range, median and mode from a grouped frequency table

D C

1 A machine weighs potatoes as they pass through the cleaning station in a packing factory.
The table shows the weights of some of the potatoes.

weight, w g	Frequency
$75 \leqslant w < 85$	22
$85 \leqslant w < 95$	46
$95 \leqslant w < 105$	55
$105 \leqslant w < 115$	46
$115 \leqslant w < 125$	32

a Write down the modal class.

b Estimate the range of this data.

c Which class interval contains the median?

D

2 The number of vehicles passing under a motorway bridge every hour for a period of one week was recorded. The frequency table shows the results.

Number of vehicles	0–499	500–999	1000–1499	1500–1999	2000–2499
Frequency	27	58	16	54	13

a Write down the modal class.

b Estimate the range of this data.

c Which class interval contains the median?

d A nearby road is closed for maintenance. This means that the number of vehicles passing under the bridge for the next 6 hours will be about 1750 vehicles each hour.

This data is added to the frequency table above. What effect will adding this data have on
  i the modal class
  ii the estimated range
  iii the class interval containing the median?

e Is it possible to work out an exact mean for this data? Give a reason for your answer.

C

AO2

4.5 Estimating the mean from a grouped frequency table

1 A machine weighs potatoes as they pass through the cleaning station in a packing factory. The table shows the weight of some of the potatoes.

C

Weight, w (g)	Frequency	Midpoint	Midpoint × frequency
$75 \leqslant w < 85$	22	80	$80 \times 22 = 1760$ g
$85 \leqslant w < 95$	46		
$95 \leqslant w < 105$	55		
$105 \leqslant w < 115$	46		
$115 \leqslant w < 125$	32		

a Copy and complete the table to work out an estimate of the total mass of the potatoes.

b Calculate an estimate of the mean mass of the potatoes.

2 The students in Mr O'Toole's biology class grew some green bean seedlings. Some of the seedlings were given fertiliser and some were not. The students used the tally chart below to record their data.

C

Height, h (cm)	With fertiliser	Without fertiliser
$0 \leqslant h < 20$		||
$20 \leqslant h < 40$	|	|||
$40 \leqslant h < 60$	~~||||~~	~~||||~~ |||
$60 \leqslant h < 80$	~~||||~~ ~~||||~~ ~~||||~~ ||	~~||||~~ ~~||||~~ ~~||||~~ ~~||||~~ |
$80 \leqslant h < 100$	~~||||~~ ~~||||~~ |||	||||
$100 \leqslant h < 120$	~~||||~~ |	|

a Will Mr O'Toole's students be able to use the tally chart to calculate an exact mean height for the green beans grown without fertiliser?
Give a reason for your answer.

b Draw a frequency table for the 'without fertiliser' results. Use it to calculate an estimate of the mean height of those seedlings. Give your answer to one decimal place.

c Calculate an estimate of the mean height of all of the green bean seedlings.
Give your answer to one decimal place.

d One of Mr O'Toole's students wrote this conclusion.
Do you agree with this student?
Give reasons for your answer.

> Conclusion
> The without fertiliser results were below average. This means that the green beans grown with fertiliser are taller, stronger and healthier than those grown without.

AO2

5 Probability

Links to:
Middle Student Book
Ch5, pp.73–94

Key Points

Calculating the probability an event does not happen **E**

$\left(\text{Probability that an event will not happen}\right) = 1 - \left(\text{Probability that an event will happen}\right)$

Mutually exclusive events **D**

Mutually exclusive events cannot happen at the same time.

P(A or B) = P(A) + P(B)

Two-way tables **E D**

A two-way table shows two or more sets of data at the same time.

The number of times an event is likely to happen **D**

Expected frequency = probability of the event happening × number of trials

Calculating relative frequency **C**

Relative frequency is also known as experimental or estimated probability.

$$\text{Relative frequency} = \frac{\text{number of successful trials}}{\text{total number of trials}}$$

Independent events **C**

Two events are independent if the outcome of one does not affect the outcome of the other.

P(A and B) = P(A) × P(B)

Drawing tree diagrams **B**

The probability of each outcome is written on the branch of the tree.

To calculate the probability of combined outcomes, multiply the probabilities.

This tree diagram is for flipping a coin twice.

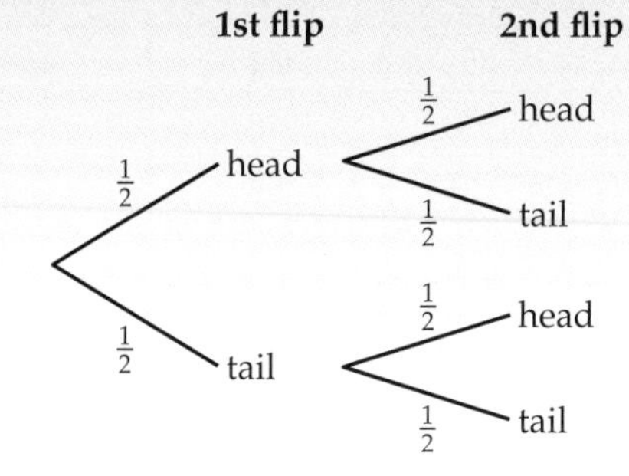

P(H and H) = $\frac{1}{2} \times \frac{1}{2} = \frac{1}{4}$

5.1 Probability that an event does not happen

E

1 The probability of picking the ace of diamonds from a pack of cards is $\frac{1}{52}$. What is the probability of **not** picking the ace of diamonds from a pack of cards?

2 The probability of winning a prize in the UK Thunderball Lottery is 0.055. What is the probability of **not** winning a prize in the UK Thunderball Lottery?

3 Simon is learning to play darts.
The probability that he hits the dartboard is 1 in 4.
What is the probability that his next dart

a hits the dartboard

b doesn't hit the dartboard?

4 The probability that this spinner lands on A is 0.6.
The probability that it lands on blue is 90%.
What is the probability that

a the spinner does not land on A?
Give your answer as a percentage.

b the spinner does not land on blue?
Give your answer as a fraction.

c the spinner lands on C?
Give your answer as a decimal.

5 Alan buys a special six-sided spinner.
The spinner is numbered 1 to 6.
The probabilities of different scores are listed in the table.

Number	1	2	3	4	5
Probability	0.35	0.05	0.1	0.2	0.3

a Work out the probability of
  i not getting a 5
  ii not getting a 2
  iii not getting a 6.

b Explain what your answer to part a iii means.

E

6 Megan has a biased four-sided dice numbered 1 to 4.
The probability of getting a 1 with this dice is $\frac{1}{2}$.
Megan says, 'There are three other numbers. So the probability of not getting a 1 with this dice is $\frac{3}{4}$.'
Explain why Megan is wrong.

A02

7 A tube of jelly beans contains 10 jelly beans. Leanne picks one at random.
The probability Leanne doesn't pick a blueberry flavour jelly bean is $\frac{3}{5}$.
How many blueberry flavour jelly beans does Leanne have in the tube?

E A03

5.2 Mutually exclusive events

1 A bag contains 20 jelly beans. Six of the jelly beans are caramel flavour, six are lime flavour, five are strawberry flavour and three are baked bean flavour.
One jelly bean is taken from the box at random.
What is the probability that the jelly bean

a is not baked bean flavour
b is not caramel flavour
c is caramel or lime flavour
d is caramel or strawberry flavour
e is caramel or baked bean flavour
f is not caramel or baked bean flavour
g is not lime or baked bean or caramel flavour?

E D

2 Work out the probability of rolling a 2 or a 4 or a 6 with a fair six-sided dice.

D

3 An art box contains four different types of pens.
One pen is taken from the box at random.
The table shows the probabilities of taking each type of pen.

Pen	Probability
ink	0.2
gel	
waterproof	0.35
glitter	0.15

a What is the probability that the pen is an ink or a waterproof?

b What is the probability that the pen is a gel?

D

4 A box contains four types of biscuits.
One biscuit is taken from the box at random.
The table shows the probabilities of taking each type of biscuit.

Biscuit	Probability
Rich tea	0.2
Choc-chip	0.2
Ginger	
Digestive	

AO2

There are five times as many ginger biscuits as digestive biscuits.
What is the probability that the biscuit is a digestive?

D

5 Einir puts 12 CDs into a bag.
Lewis puts eight computer games into the same bag.
Karl puts some DVDs into the bag.
The probability of taking a DVD from the bag at random is $\frac{1}{5}$.
How many DVDs did Karl put in the bag?

AO3

5.3 Two-way tables

E

1 The table shows the number of students at Oakwood School who do or don't have a dog.

	Dog	No dog	Total
Girls	86	36	122
Boys	47	91	138
Total	133	127	260

One student is chosen at random. What is the probability that this student

a is a girl who has a dog

b is a boy who has a dog

c doesn't have a dog?

E

2 The table shows the age and gender of a sample of 40 teenagers at a theme park.

	Age in years							
	13	14	15	16	17	18	19	Total
Boys	2	1	3	8	6	2	1	23
Girls	0	3	5	2	4	1	2	17
Total	2	4	8	10	10	3	3	40

D

A teenager is chosen at random. What is the probability that this teenager

a is a 15-year-old girl

b is an 18-year-old boy

c is 13 years old

d is a boy

e is less than 17 years old

f is not 16 years old?

On one particular day there are 2000 teenagers at the theme park altogether.
How many of these are likely to be

g 14 years old

h girls?

3 Copy the two-way table to show the numbers of girls and boys in class 10G who are left-handed or right-handed.
Use the information given to work out the missing numbers in the table.

	Left-handed	Right-handed	Total
Girls	?	?	?
Boys	?	?	?
Total	?	?	30

The probability that a boy is right-handed is $\frac{13}{30}$.
The probability that a pupil is left-handed is $\frac{1}{6}$.
The probability that a pupil is a girl is $\frac{7}{15}$.

D

AO2

5.4 Expectation

D

1 A fair four-sided dice is rolled 80 times.

How many times would you expect it to land on:

a the number 3 **b** a number less than 4?

2 Alice has these number cards.

1	2	3	4	5	6	7	8	9	10

She shuffles the cards and selects one at random.
She puts the card back in the pack.
She does this 200 times.

How many times would you expect her to select

a the number 10 card

b an even number card

c a square number card

d a prime number card?

D

3 Steffan has a bag of music CDs.
In the bag there are six rock, four jazz, two classical and three pop music CDs.
Steffan selects one CD at random from the bag and then replaces it.
He does this 90 times.
How many times would you expect him to select a classical CD?

4 At a school fête, Robyn runs a 'Wheel of fortune' game.
She charges £1 to spin the wheel.
The wheel is equally likely to stop on any number.
If the wheel stops on a square number she gives a prize of £2.
Altogether 200 people play the game.
How much money would you expect Robyn to make for her school?

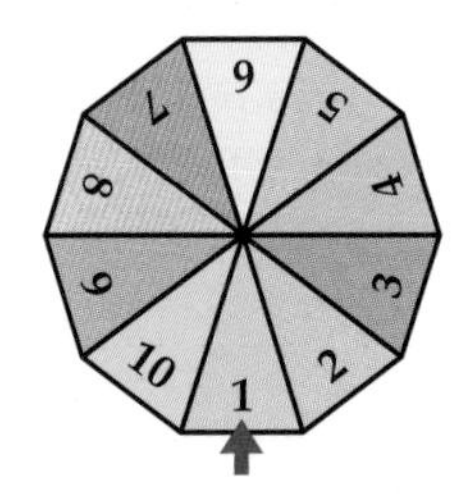

5 This table shows the probability of selecting coloured counters from a bag.

Colour	Red	Blue	Green	White	Yellow
Probability	0.3	0.1	0.25	?	?

The probability of selecting red is twice the probability of selecting white.
Lubna selects one counter at random from the bag and then replaces it.
She does this 300 times.
How many times would you expect her to select a yellow counter?

AO2

5.5 Relative frequency

C

1 Philip rolls a normal six-sided dice.
He keeps a tally of how many 5s he rolls.
The table shows his results.

Number of rolls	20	50	100	200	500	1000
Number of 5s	2	6	13	28	75	170
Relative frequency						

a Calculate the relative frequency for the number of 5s at each stage of the experiment.

b Work out the theoretical probability of obtaining a 5.

AO2

c Do you think the dice is fair? Explain your answer.

C

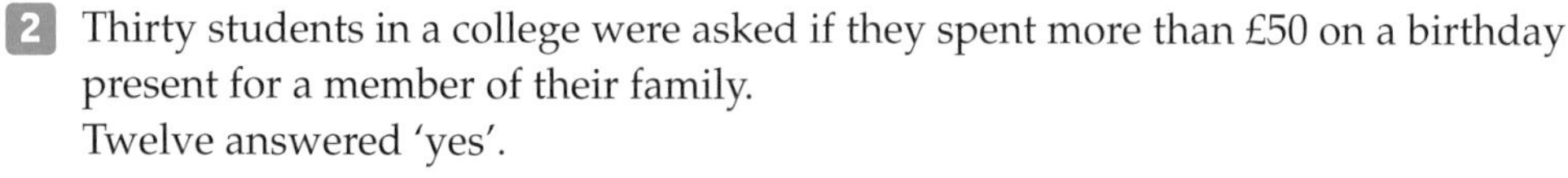

2 Thirty students in a college were asked if they spent more than £50 on a birthday present for a member of their family.
Twelve answered 'yes'.

a What is the relative frequency of 'yes' answers?

b There are 900 students in the college.
How many of these do you estimate will have spent more than £50 on a birthday present for a member of their family?

C

3 Sally has a spinner with three equal sections labelled 1, 2 and 3.
She spins the spinner 30 times.
Here are her results.

3 2 1 1 1 3 3 2 2 2 1 1 1 2 1

1 1 3 3 2 2 3 1 2 1 2 1 3 1 1

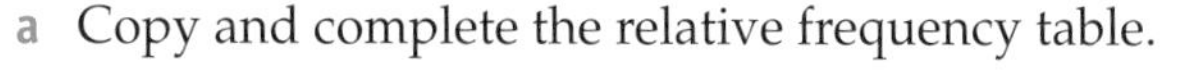

a Copy and complete the relative frequency table.

Number	1	2	3
Relative frequency			

b Sally thinks that the spinner is biased.
Write down the number you think the spinner is biased towards.
Explain your answer.

AO3

c What could Sally do to make her results more reliable?

5.6 Independent events

C

1 Bag A contains three red and five blue balls.
Bag B contains four red and six blue balls.
Priya takes one ball at random from each bag.
What is the probability that both balls are

a red b blue?

2 A fair four-sided dice is rolled twice.

a What is the probability of getting a 1 and then a 3?

b What is the probability of getting two 4s?

C

3 Erik has a fair six-sided dice and a fair spinner numbered 1 to 4.
He rolls the dice and spins the spinner at the same time.
He multiplies the number on the dice and the number on the spinner to give the score.

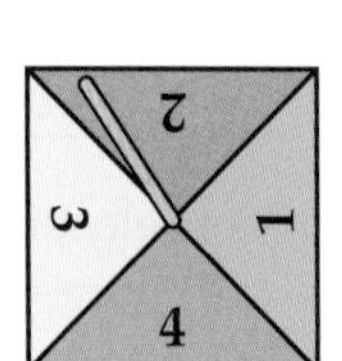

What is the probability that he gets a score

AO2

a of 9 b of 12 c greater than 12?

4 Jenica rolls a fair dice numbered 1 to 8.
Latika rolls a fair dice numbered from 1 to 4.
Work out the probability that

- a they both obtain a 4
- b they both obtain an odd number
- c the total of their scores is 4
- d Jenica's score is twice Latika's score
- e Latika's score is greater than Jenica's score.

C AO2

5.7 Tree diagrams

B

1 Draw tree diagrams to show all the possible outcomes in each part.
Label the branches with the appropriate outcomes.

- a For lunch today I have a choice of a salad or a baguette.
  For lunch tomorrow I have the same choice.
- b A bag contains red and blue balls.
  I take a ball at random, record its colour, then put it back in the bag.
  I take a second ball at random, record its colour, then put it back in the bag.
- c For dinner tonight I can have either pasta, risotto or omelette, followed by either ice cream or fruit.

2 Sandra has a bag containing 4 yellow counters and 5 blue counters.
She takes a counter at random from the bag, records the colour then puts it back in the bag.
She then takes a second counter from the bag.

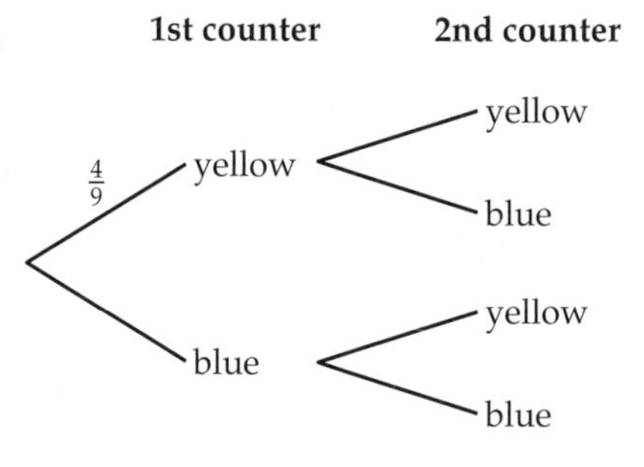

- a Copy and complete the tree diagram to show all the possible outcomes and their probabilities.
- b Work out the probability that Sandra takes
  - i two yellow counters
  - ii two blue counters
  - iii a yellow then a blue counter
  - iv a blue then a yellow counter.

B

3 Lee, Jin and Tao all go to the same school.
The probability that Lee brings a packed lunch is 0.2.
The probability that Jin brings a packed lunch is 0.1.
The probability that Tao brings a packed lunch is 0.5.

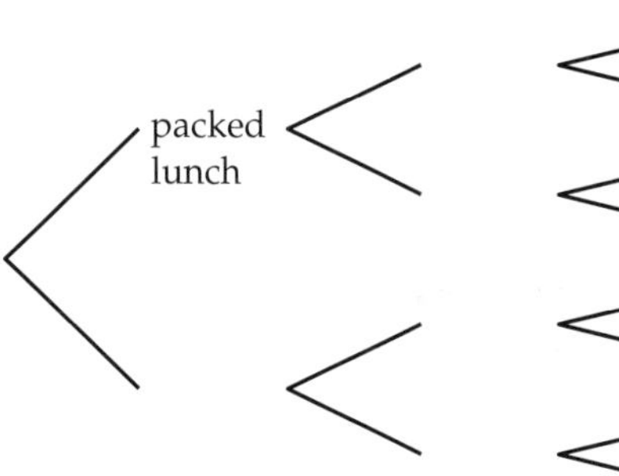

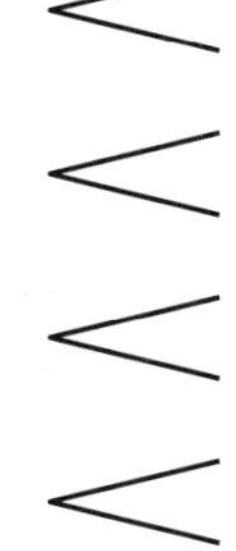

- a Copy and complete the tree diagram to show all possible outcomes.
- b On any day, what is the probability that
  - i all three bring a packed lunch
  - ii all three don't bring a packed lunch
  - iii Lee brings a packed lunch but Jin and Tao don't
  - iv Jin and Tao bring a packed lunch and Lee doesn't?

AO2

6 Cumulative frequency

Links to:
Middle Student Book
Ch6, pp.95–108

Key Points

Drawing and using a cumulative frequency diagram for grouped data

B

Cumulative frequency is a running total of the frequencies.

You can use a cumulative frequency diagram to estimate the median $\left(\frac{n}{2}\text{th value}\right)$, lower quartile $\left(\frac{n}{4}\text{th value}\right)$ and upper quartile $\left(\frac{3n}{4}\text{th value}\right)$. The inter-quartile range indicates the spread of the data.

Inter-quartile range = upper quartile − lower quartile

Drawing a box plot from a cumulative frequency diagram

B

A box plot shows the range, lower and upper quartiles, and median of a set of data.

Draw the box plot on the same scale as the cumulative frequency diagram from which the data comes.

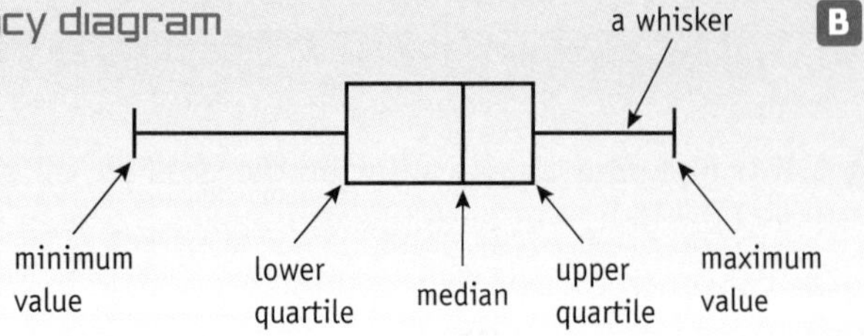

Comparing data sets and drawing conclusions

B

The median is one of the measures of 'average'. You should always compare the medians of the data sets you are looking at. The inter-quartile range is a measure of consistency and tells you the spread of the middle 50% of the data. Always comment on the sizes of the inter-quartile range of the distributions.

6.1 Cumulative frequency

B

1 Imir sells electrical goods on the internet. He posts the goods to his customers. The cost of postage depends on the weight of the parcel. This table shows the weights of the parcels Imir posted last month.

Weight, w (kg)	$0 < w \leqslant 2$	$2 < w \leqslant 4$	$4 < w \leqslant 6$	$6 < w \leqslant 8$	$8 < w \leqslant 10$	$10 < w \leqslant 12$	$12 < w \leqslant 14$	$14 < w \leqslant 16$
Frequency	3	18	14	9	6	5	12	10

a Draw a cumulative frequency diagram to illustrate this data.

b Use the cumulative frequency diagram to estimate

i the median

ii the lower quartile

iii the upper quartile

iv the inter-quartile range.

c Parcels up to 3 kg in mass cost £5.20 each to post. Estimate the cost of posting all the parcels weighing less than 3 kg.

d Parcels weighing between 5.75 kg and 8.25 kg cost £10.25 each to post. Estimate the cost of posting all the parcels weighing between 5.75 kg and 8.25 kg.

AO2

6.2 Box plots

B

1 For each of the box plots shown, work out

a the median

b the upper quartile

c the inter-quartile range

d the largest possible data value

e the range.

Box A

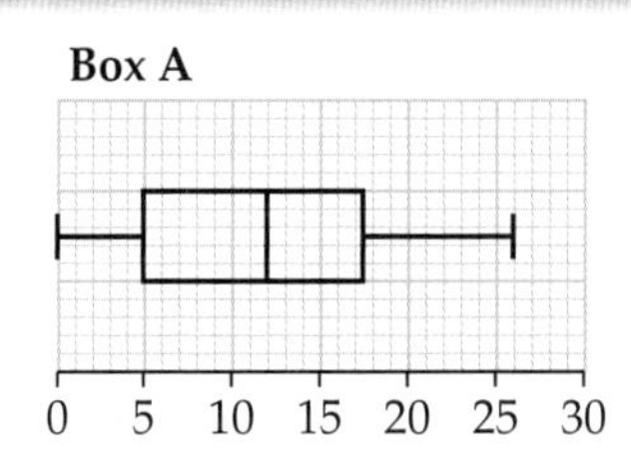

Box B

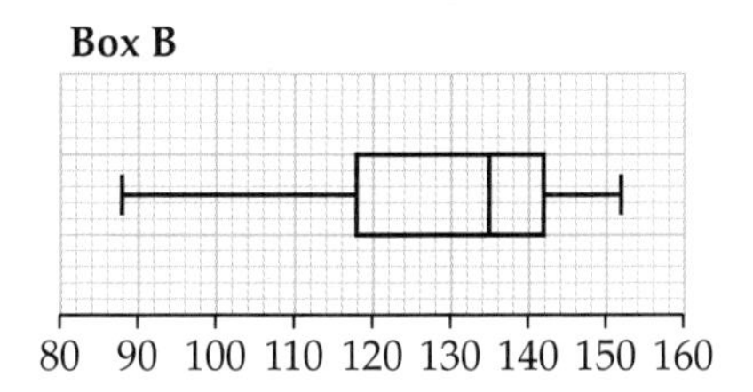

2 A quality control officer at a flour mill weighed 80 bags of flour. Each bag of flour should have weighed 1 kg.
The cumulative frequency diagram shows the actual weights of the bags.

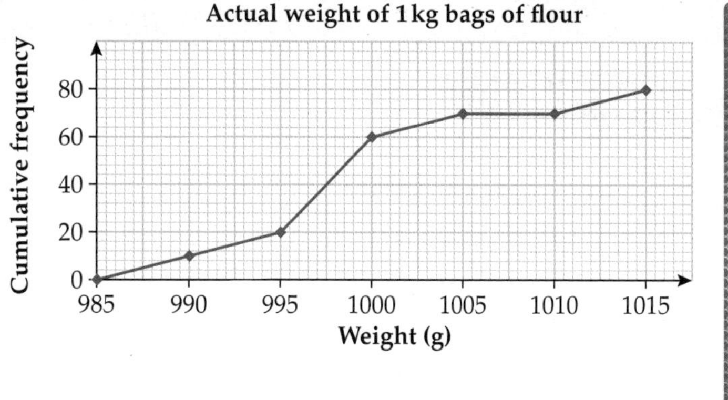

a How many of the bags contained between 995 g and 1005 g of flour?

b Use the cumulative frequency diagram to estimate

i the median weight

ii the inter-quartile range.

c Draw a box plot to represent this data.

d Estimate the range of weights for the lightest 25% of the bags of flour.

e Estimate the range of weights for the heaviest 50% of the bags of flour.

B

AO2

6.3 Comparing data sets and drawing conclusions

1 The box plots show the length of time taken for 40 boys and 60 girls to complete a puzzle.

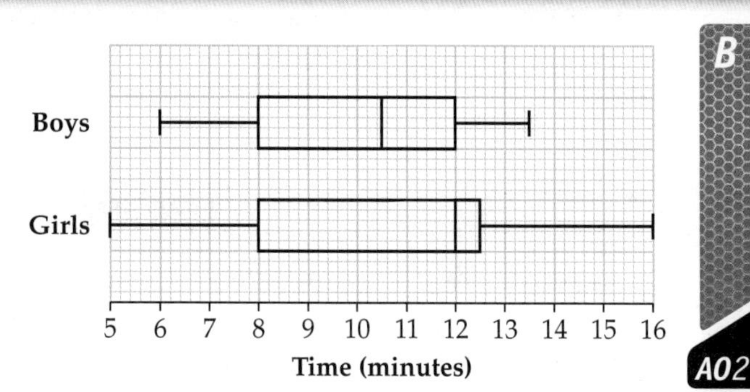

a What is the median time taken by the girls?

b What was the quickest time taken by the boys?

c Write down two differences between the times taken by the girls and the boys.

d How many students completed the puzzle in less than 12 minutes?

B

AO2

2 Mike measured the height gain of two different varieties of hedge plants. The frequency distributions of each variety are shown on the cumulative frequency diagram.

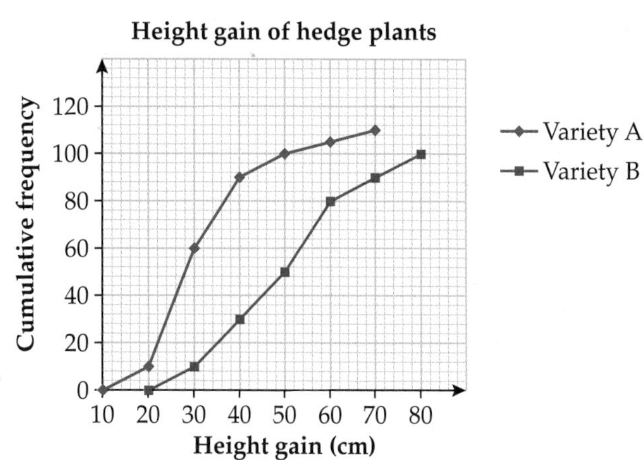

a How many hedge plants of variety A were measured?

b Estimate the minimum and maximum height gains for variety A.

c Write down the median for variety A.

d Work out the inter-quartile range for variety A.

e How many hedge plants of variety B were measured?

f Estimate the minimum and maximum height gains for variety B.

g Write down the median for variety B.

h Work out the inter-quartile range for variety B.

i Draw box plots for both varieties.
Use a scale of 0 to 90 on the horizontal axis. Draw one box plot directly beneath the other.

j Compare the two varieties, giving reasons for any statements you make.

B

AO3

Ratio and proportion

Links to:
Middle Student Book
Ch7, pp.109–129

Key Points

Simplifying ratios E

To simplify a ratio, divide each of the numbers in the ratio by their highest common factor. If the values have units, such as kg or g, make sure that all the quantities are in the same units.

Writing a ratio as a fraction D C

The numerator is the part of the ratio you want to write as a fraction. The denominator is the total number of parts of the ratio.

Writing ratios in the form $1:n$ or $n:1$ C

First decide which number in the ratio you want to be 1, then divide all the numbers in the ratio by that number.

Divide a quantity in a given ratio D C

To share in a given ratio

- work out the total number of parts to share into
- work out the value of one part
- work out the value of each share by multiplying the value of one part by the numer of parts in that share.

Direct proportion D

When two values are in direct proportion

- if one value is zero, so is the other
- if one value doubles, so does the other.

Unitary method D

To solve proportion problems using the unitary method, work out the value of one unit first.

Best buys D

The best buy means the product that gives you the best value for money. To compare two prices and sizes, work out the price for one unit of each and compare those.

Exchange rates D C

The exchange rate tells you how many euros, or any other currency, you can buy for £1. You can use the unitary method to convert between currencies. First find the value of one unit of the currency.

Inverse proportion C B

When two values are in inverse proportion, one increases at the same rate as the other decreases.

7.1 Ratio

E

1 Simplify each of these ratios.

a $9:3$ b $6:18$ c $40:100$ d $75:50$
e $120:130$ f $900:1000$ g $7:56$ h $81:9$

2 Simplify each of these ratios.

a 5 mm : 3 cm b 2 kg : 700 g c 4.4 kg : 0.2 kg d 2.4 m : 8 cm

3 Simplify each of these ratios.

a $\frac{1}{2}:2$ b $10:2\frac{1}{2}$ c $1\frac{1}{4}:5$ d $3.5:5.5$

4 Which of these ratios is equivalent to $3:4$?

A $12:9$ **B** $12:16$ **C** $16:20$ **D** $20:60$

E

5 Bianca said that the ratio $25:20$ simplifies to $4:5$.
Is Bianca correct?
Give a reason for your answer.

A02

6 George makes up orange squash using 50 ml of concentrate with half a litre of water.
Write the ratio of concentrate to water in its simplest form.

7 Here is a recipe for salmon fishcakes.
Dai has 1.25 kg of salmon fillets and wants to use them all in some fishcakes.

a How much of each other ingredient does Dai need to make the fishcakes?

b How many people can Dai serve?

Salmon fishcakes (serves 4)
500 g salmon fillets
1 tbsp olive oil
400 g potatoes
2 large eggs

7.2 Ratios and fractions

1 Sharon and Kate set off some fireworks. The ratio of Sharon's fireworks to Kate's fireworks is 2 : 3.

a What fraction of the fireworks does Sharon set off?

b What fraction of the fireworks does Kate set off?

2 Chin and Damita share a chocolate cake in the ratio 3 : 7.

a What fraction of the chocolate cake does Chin eat?

b What fraction of the chocolate cake does Damita eat?

3 The ratio of divers to swimmers training at a swimming pool is 2 : 9.
Esta says, 'Two-ninths of the people training are divers.' Is Esta correct?
Explain your answer.

4 A recipe for mango and pistachio fool uses mango and pistachios in the ratio 25 : 3 by weight. Frank uses 575 g of mango. What weight of pistachios should he use?

5 In a pond the ratio of goldfish to koi carp is 2 : 5. There are 16 goldfish in the pond. How many fish are there in the pond altogether?

7.3 Ratios in the form 1 : n or n : 1

1 Write each of the following ratios in the form

i 1 : n and ii n : 1.

a 4 : 10 b 6 : 4 c 5 : 8 d 10 : 9

e £3.60 : £4.80 f 2 m : 5 cm g 10 years : 6 months h 8 g : 5 kg

2 The ratio of shiraz grape to pinot noir grape in a bottle of French wine is 184 : 21.

a Write this as a ratio in the form n : 1.

The ratio of shiraz grape to pinot noir grape in a bottle of Spanish wine is 59 : 7.

b Write this as a ratio in the form n : 1.

c Compare your answers to **a** and **b**. Which wine has the higher proportion of shiraz grape? Explain your answer.

3 The ratio of fat : flour : sugar in one type of biscuit is 4 : 5 : 3.
The ratio of fat : flour : sugar in another type of biscuit is 12 : 11 : 8.
Which biscuit, the first or the second, has the higher proportion of sugar? Show working to support your answer.

7.4 Dividing quantities in a given ratio

1 Share £30 in the ratio 2 : 1.

2 Share £30 in the ratio 3 : 1.

3 Share £30 in the ratio 4 : 1.

4 Share £30 in the ratio 5 : 1.

5 Share £30 in the ratio 2 : 13.

6 Share £30 in the ratio 1 : 2 : 3.

7 Ruth and Gordon buy a bag of sweets for 70p.
Ruth pays 20p and Gordon pays 50p.

- a Write the amount they pay as a ratio.
- b There are 21 sweets in the bag. Work out how many sweets Gordon should have.

8 The ratio of pupils to teachers to support staff on a school trip is 25 : 2 : 1.
There are 840 people on the trip altogether. How many of them are pupils?

9 Jafar makes purple paint by mixing red, white and blue paint in the ratio 3 : 10 : 5. He makes 9 litres of purple paint. How much of each colour does he use?

AO2

7.5 More ratios

1 Harriet makes a fruit cake. She uses 150 g of dried fruit for every 250 g of flour.

- a Write these amounts of dried fruit and flour as a ratio.
- b Simplify your ratio as much as possible.
- c How much flour does she need to go with 90 g of dried fruit?
- d How much dried fruit does she need to go with 350 g of flour?

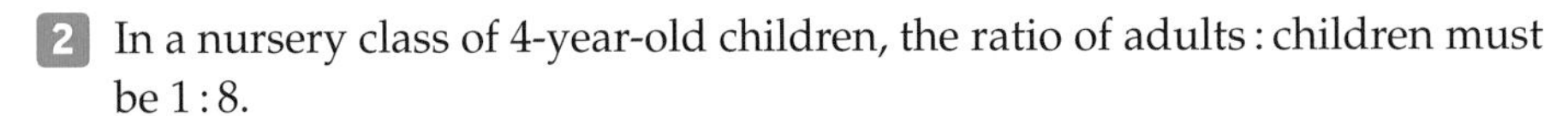

2 In a nursery class of 4-year-old children, the ratio of adults : children must be 1 : 8.

- a How many children can 3 adults look after?
- b How many adults are needed to look after 32 children?

3 Inga's favourite drink is grenadine and lemonade mixed in the ratio 2 : 11.

- a How much lemonade does she need to mix with 30 m*l* of grenadine?
- b How much grenadine does she need to mix with 1 litre of lemonade? Give your answer to the nearest m*l*.

AO2

4 The table shows the staff : customer ratio for different ages of customers riding at a stables.

Age of customer	Staff : customer ratio
4–7 years	1 : 1
8–16 years	1 : 3
17 years+	1 : 5

This is the booking information for one ride:

2 pm beach ride – $1\frac{1}{2}$ hours

8 4 – 7 years old
11 8 – 16 years old
18 17 years +

Work out the minimum number of staff needed to take out this ride.

C

AO2

7.6 Proportion

1 Jabilo works 5 hours a day in a shop and gets paid £37.50. One day he is asked to work 7 hours. How much will he get paid for this day?

2 A group of 6 people go horse riding. Altogether they pay £225. How much does it cost 2 people to go horse riding?

3 Sixteen caramel biscuits weigh 464 g. How much do 5 caramel biscuits weigh?

4 Nine chocolate bars contain 963 calories. How many calories are there in 2 chocolate bars?

5 A 300 g bag of tortilla chips contains 1488 calories. How many calories are there in a 25 g serving?

6 The average number of beans in a can of baked beans is 549.
A can of beans contains 366 calories. How many calories are there in a spoonful of 21 baked beans?

7 Five friends plan to go to a concert. The tickets will cost a total of £168.75.
Another friend decides he wants to go as well.
How much will the six tickets cost?

D

AO2

7.7 Best buys

1 A multipack of 5 muesli bars costs £1.25. How much does one muesli bar cost?

2 A 1.5 litre carton of banana smoothie costs £4. How much banana smoothie do you get for £1?

E

3 Kiyoshi is paid £817 for working 38 hours a week. Lars is paid £2793 for working 140 hours a month. Who gets paid the higher wage per hour? Show your working.

4 Here are two bottles of water. Which is the better buy? Explain your answer.

5 Here are three cartons of apple juice. Which is the best value for money? Explain your answer.

7.8 More proportion problems

Use the following information to answer Q1 and Q2.

Exchange rate for pounds (£) to euros (€) is £1 = €1.14

1 **a** Convert £228 to euros.

b Convert €228 to pounds.

2 Lianne is in Spain on holiday. She sees a mobile phone in a shop on sale for €429.99. Lianne remembers seeing the same phone on sale in the UK for £369.99. Where should Lianne buy the mobile phone? Explain your answer.

3 It takes 3 plumbers 2 days to renew the plumbing in a house. How long will it take 5 plumbers to renew the plumbing in the same house?

4 A vase manufacturer has an order for 756 vases. The owner knows that one person can pack 35 vases in one hour. She wants all of the vases packed in one shift of 8 hours. How many people are required to do the job?

5 An office block has 92 identical offices. The cleaning company allow an hour and a half for one person to clean 6 offices. All the offices must be cleaned between 8 pm and 12 midnight. How many cleaners need to be employed to clean the offices?

8 Complex calculations

Links to:
Middle Student Book
Ch8, pp.132–142

Key Points

Repeated percentage change C

When you invest money the interest is usually added to the amount of money originally invested. The interest that you get is calculated on the amount invested plus the interest already received. This is known as compound interest.

1 Add the rate of interest to 100%.
2 Convert this percentage to a decimal to get the multiplier.
3 Multiply the original amount by the multiplier as many times as the number of years for which the money is invested.

You can also use this method to work out a repeated percentage loss or depreciation. To find the multiplier for loss or depreciation you must first subtract the percentage from 100%.

Reverse percentages B

To work out the original quantity when you are given the quantity after a percentage increase or decrease, use one of these methods.

Method A

1 Work out what percentage the final quantity represents.
2 Divide by this percentage to find 1%.
3 Multiply by 100 to get the 100% figure.

Method B

1 Work out what percentage the final quantity represents.
2 Divide by 100 to get the multiplier.
3 Divide the final quantity by the multiplier.

Using and interpreting standard form B

A number in standard form is a number between 1 and 10 ($1 \leq n < 10$) multiplied by an integer power of 10.

8.1 Repeated percentage change

C

1 Mesha buys a car for £17 995.
He reads that the car will depreciate by 12% a year.
How much will the car be worth at the end of 3 years?
Give your answer to the nearest £100.

2 Tyra invests £7500 in a 5-year government bond.
The bond pays compound interest at a rate of 3.65% per annum.
How much will her investment be worth at the end of the 5 years?

C

3 Three years ago Rico put £3000 into an investment account. The account paid compound interest at a rate of 7% per annum.
He decides to leave the money in the account for a further 2 years. The new rate of compound interest is 5% per annum.
How much will his original investment have grown by in 2 years' time?

4 How many years will it take for an investment of £10 to reach £20 at a compound interest rate of 4.5% per annum?

5 Sam invested £1000. At the end of 4 years her investment was worth £1272.
What was the annual rate of compound interest that Sam received?

AO2

8.2 Reverse percentages

B

1 After a pay rise of 2.9% Emyr earns £23 049.60 per year.
What did he earn before his pay rise?

2 Bill bought a coat in the sale.
He paid £46.75.
What was the price of the coat before the sale?

B

3 In a beach clean 62% of the items of litter collected were plastic.
The remaining 1425 items collected were not plastic.
How many plastic items were collected?

4 Tao has paid a 20% deposit on a moped.
He still has £2076 to pay.
His insurance works out as 9% of the full cost of the moped.
How much is Tao's insurance?

AO2

5 Sally buys a new laptop for her business. She pays £423, including VAT.
Sally can get a refund on the VAT. The rate of VAT is 17.5%.
How much will Sally get as a refund?

8.3 Standard form

B

1 Write the following numbers in standard form.

a 7000 b 700 000 c 7200 d 720 000

e 0.007 f 0.000 07 g 0.007 89 h 0.007 08

2 Write the following as decimal numbers.

a 6×10^4 b 6×10^{-4} c 3.81×10^7 d 3.81×10^{-1}

3 Write the following numbers in standard form.

a 16.3×10^2 b 0.04×10^3

c 12.9×10^{-4} d 0.003×10^{-6}

e 3.5 million f $\frac{7}{8}$

g $27 \times 10^6 \times 0.005$ h $\sqrt{20 \times 20 \times 20 \times 20}$

4 Use your calculator to work out the following, giving your answers in standard form.

a $3.6 \times 10^4 \times 5 \times 10^2$ b $(7.2 \times 10^{-4}) \div (1.2 \times 10^6)$

B

AO2

5 Light from the Sun takes approximately 8 minutes and 20 seconds to reach Earth.
Light travels at 3×10^8 metres per second.
Estimate the distance of the Earth from the Sun in km.

9 Number skills

Links to:
Middle Student Book
Ch9, pp.147–157

Key Points

Multiplying whole numbers (E D)

To multiply whole numbers, use the grid method or the standard column method.

Dividing using repeated subtraction (E D)

You can use repeated subtraction for division. The number you divide by is called the divisor. Keep subtracting multiples of the divisor until you cannot subtract any more. Then see how many lots of the divisor you have subtracted altogether.

If the answer to a division is not exact, there is a remainder.

If you are answering a word problem, you may need to decide whether to round the answer to the division up or down.

Estimating (E D C)

You can use estimation to check that an answer is about right.

To estimate
- round all the numbers to one significant figure
- do the calculation using these approximations.

Multiplying and dividing negative numbers (E)

When multiplying or dividing two numbers you need to check the signs.
- If the signs are the same, the answer is positive.
- If the signs are different, the answer is negative.

Multiplication	Division
$+ \times + = +$	$+ \div + = +$
$+ \times - = -$	$+ \div - = -$
$- \times + = -$	$- \div + = -$
$- \times - = +$	$- \div - = +$

9.1 Multiplying and dividing whole numbers

1 Lyn organises a holiday to Slovenia for herself and 11 friends. The total cost of the holiday is £975 per person. The travel company wants a deposit of £245 per person. Lyn sends a cheque for the total deposit to the travel company.

a What is the total deposit for the 12 friends?

The remaining money is to be paid at the start of the holiday.

b What is the total amount that the friends must pay at the start of the holiday?

E AO2

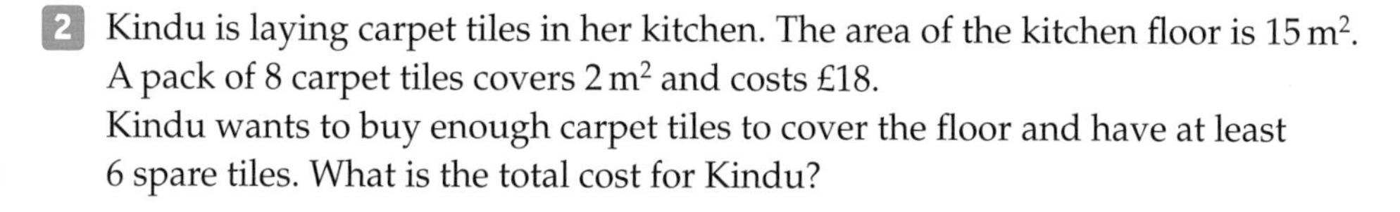

2 Kindu is laying carpet tiles in her kitchen. The area of the kitchen floor is $15\,m^2$.
A pack of 8 carpet tiles covers $2\,m^2$ and costs £18.
Kindu wants to buy enough carpet tiles to cover the floor and have at least 6 spare tiles. What is the total cost for Kindu?

3 You are given that $27 \times 152 = 4104$. What is the value of

a 270×152 **b** $41\,040 \div 27$ **c** $4104 \div 15.2$ **d** $4104 \div 1520$?

4 At the start of January, a tile shop receives a delivery of 14 boxes of tiles.
Each box contains 48 tiles. All the tiles are the same design.
One out of every 24 tiles is damaged and is thrown away.
On average the tile shop sells 60 of this design of tile per month.
During which month will the shop run out of these tiles?

9.2 Estimation

1 Work out an approximate value for each of these calculations.

a 7.2×9.4 b 45×3.7 c $84 \div 19$ d $683 \div 12$

Remember to round all of the numbers to one significant figure.

D

2 Estimate an answer to each of these calculations.

a $\frac{678 + 313}{18}$ b $\frac{442 \times 3.78}{7.81}$ c $\frac{28.45 + 74.2}{9.93 - 5.7}$

C

3 Estimate an answer to each of these calculations.

a $\frac{47.332 \times 395.8}{1.963 \times 23.69}$ b $\frac{845.9 + 78.609}{10.5956}$ c $\frac{320.896}{24.887 \times (12.09 - 9.951)}$

C

4 A 5-litre can of decking oil will cover an area of about 28.5 m² of decking.
Joe wants to put three coats of decking oil on an area of 115 m².
Which is the best approximate calculation to use to work out how many litres of decking oil Joe needs?

A $\frac{3 \times 100 \times 5}{30}$ B $\frac{3 \times 100 \times 5}{20}$ C $\frac{3 \times 120 \times 5}{30}$ D $\frac{3 \times 120 \times 5}{20}$

AO2

Give a full reason for your answer.

9.3 Negative numbers

E

1 Work out

a $3 \times (-4)$ b $(-3) \times (+6)$ c $16 \div (-4)$ d $(-42) \div (-6)$

2 Copy and complete these calculations.

a $4 \times \square = -24$ b $\square \times 9 = -45$

c $-100 \div \square = 20$ d $\square \times (-7) = -63$

E AO2

3 Here is a sequence of numbers.

7, 2, −3, −8, ...

Work out the next three numbers in the sequence.

E AO3

4 The sum of two numbers is −4.
When they are divided the answer is −2.
What are the two numbers?

D

5 Work out

a $(-4) \times 3 \times (-5)$ b $(-6) \times (-2) \times (-4)$ c $(-22) \div (-11) \div (-1)$

6 Copy and complete these calculations.

a $6 \times (-2) \times \square = -60$ b $(-3) \times \square \times 5 = 45$ c $100 \div \square \div (-2) = -25$

D AO3

7 You are given that

$a + b = 0$

$a + c = -5$

$a \times b \times c = 12$

a and c are negative numbers. b is a positive number.
All the numbers are different.
What are the values of a, b and c?

10 Factors, powers and standard form

Links to:
Middle Student Book
Ch10, pp.158–178

Key Points

Finding multiples E

When you multiply a number by any whole number, you get a multiple of the first number.

Prime numbers E

A prime number is a number with exactly two factors: 1 and the number itself.

Factors and prime factors E

The factors of a number are the numbers that divide into it exactly. A factor that is also a prime number is called a prime factor.

Finding the LCM C

The lowest common multiple (LCM) of two numbers is the smallest number that is a multiple of both numbers.

Lowest common multiple is also known as least common multiple.

Finding the HCF C

The highest common factor (HCF) of two numbers is the largest number that is a factor of both numbers.

Finding and using prime factors C

To find a number's prime factor decomposition

- find any pair of factors of the number
- write them as 'branches' on a tree
- keep dividing the factors until you have only prime numbers at the ends of the branches.

Using squares, cubes and roots

The inverse of squaring is finding the square root ($\sqrt{\ }$). Positive numbers have two square roots: a positive square root and a negative square root.

The inverse of cubing is finding the cube root ($\sqrt[3]{\ }$).

Index notation and roots E B

You can use index notation to simplify repeated multiplication.

The number you are multiplying by itself is called the base. The index tells you how many times the base appears in the multiplication.

Any number to the power 1 is the number itself.

Any number to the power 0 is 1.

A negative power is the reciprocal of the equivalent positive power. The reciprocal of a number means 1 divided by that number.

Using prime factors to find HCFs and LCMs C

To find the HCF

1. Write each number as the product of its prime factors.
2. Find the prime factors that are common to both lists of prime factors numbers.
3. Multiply these to give the HCF.

To find the LCM

1. Write each number as the product of its prime factors.
2. For each prime factor, find the higher power in the two lists of prime factors.
3. Multiply these to give the LCM.

Multiplying and dividing powers C B

To multiply powers of the same number, add the indices.

To divide powers of the same number, subtract the indices.

Using standard form B

A number in standard form is a number between 1 and 10 ($1 \leqslant n < 10$) multiplied by an integer power of 10.

10.1 Multiples

E

1 870 is a multiple of 6. It has three digits. Write down two more multiples of 6 that have three digits.

2 Here are some number cards.

255 | 144 | 232 | 324 | 405 | 187

Which three of the numbers are not multiples of 9?

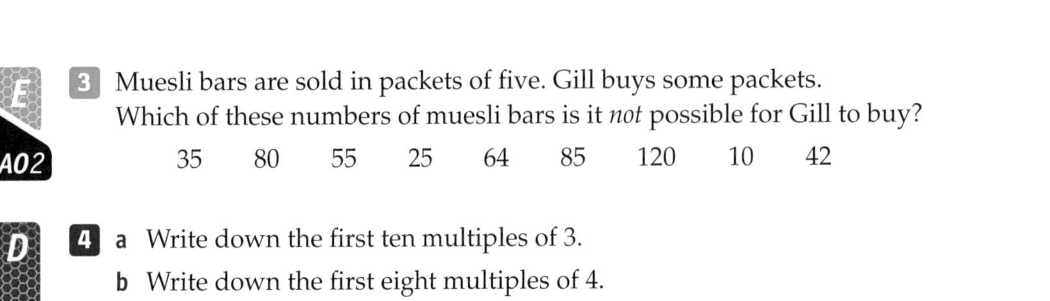

E AO2

3 Muesli bars are sold in packets of five. Gill buys some packets.
Which of these numbers of muesli bars is it *not* possible for Gill to buy?

35 80 55 25 64 85 120 10 42

D

4 **a** Write down the first ten multiples of 3.
b Write down the first eight multiples of 4.
c Write down the common multiples of 3 and 4 that are less than 30.
d What is the LCM of 3 and 4?

C

5 Work out the LCM of each of these pairs of numbers.
a 2 and 5 b 3 and 7 c 8 and 12

C

6 Davina says, 'All multiples of 4 are also multiples of 8.'
Is Davina correct? Explain your answer.

AO2

7 Carlos has some €20 notes in his pocket. He says, 'I have more than €320.'
Juan has some €50 notes in his pocket. He says, 'I have less than €600.'
They both have the same amount of money.
What are the two amounts of money that they could have?

10.2 Factors and primes

E

1 Write down true or false for each statement.
a 15 is a prime number
b 16 is a square number
c 19 is a prime number
d 27 is not a prime number
e 2 is an even prime number
f 1 is an odd prime number
g 24 is a square number
h 9 is a square prime number

E AO2

2 Dafydd is a teenager. His age is a prime number.
In four years time his age will be another prime number, but he will no longer be a teenager.
How old is Dafydd now?

E

3 a Find all the factors of 72.
b Write down all of the factors that are prime factors of 72.

E AO2

4 Rohan says, 'All factors of 16 are also factors of 32.'
Is Rohan correct? Explain your answer.

D

5 **a** Find all the factors of 48.
b Find all the factors of 60.
c Write a list of common factors of 48 and 60.
d What is the HCF of 48 and 60?

C

6 Work out the HCF of each of these pairs of numbers.
a 6 and 8 b 25 and 35 c 16 and 40.

10.3 Squares, cubes and roots

1 Work out

a 13^2

b 5^3

c the square of 15

d the cube of 7

E

2 Write down the value of

a $(-4)^2$

b $(-8)^2$

c $\sqrt[3]{27}$

d $\sqrt[3]{64}$

E

3 Copy and complete this diagram. The numbers in each pair of circles are multiplied together to give the number in the square between them.

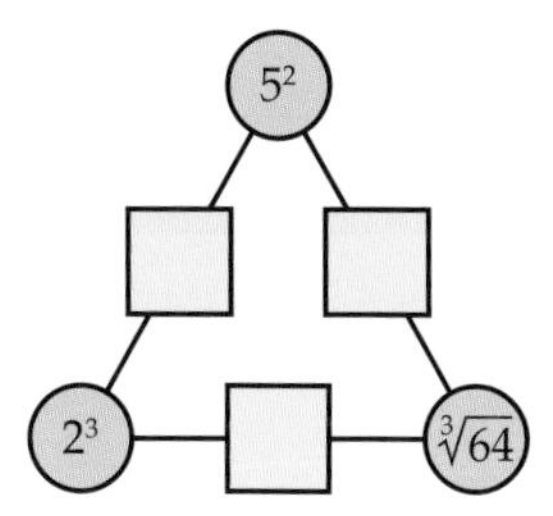

AO2

4 Write down both square roots of 64.

D

5 Work out

a $\sqrt{5^2 - 4^2}$

b $\sqrt{6^2 + 8^2}$

C

6 Write each of these as a fraction in its lowest terms.

a $\frac{2^4}{4^3}$

b $\frac{6^2}{3^4}$

7 Sandeep uses 1 cm^3 blocks to make three cubes. The first cube has a side length of 4 cm, the second cube has a side length of 5 cm and the third cube has a side length of 6 cm.
He breaks up the cubes and uses some of them to make the largest cube he can. How many blocks does he have left over?

C AO3

8 Simplify each of these.

a $\sqrt{5^2}$

b $(\sqrt{4.7})^2$

C

10.4 Indices

1 Work out

a 2^8

b 3^5

c 7^4

E

2 Three sisters have a party.
Each sister invites their three daughters.
Each daughter invites three of their friends.
Each of the friends invites three members of their family.
Assume everyone who is invited turns up to the party.

a Use index notation to write down the total number of people who are at the party.

b Calculate how many people are at the party.

E AO2

E A02

3 Which one of these number cards is the odd-one-out? Give a reason for your answer.

A	B	C	D	E
$\sqrt{8}\times\sqrt{25}$	$\dfrac{20^2}{\sqrt{64}}$	$(3^3 \times 2) - \sqrt{4}$	$\dfrac{\sqrt{100}}{2}$	$\dfrac{\sqrt[3]{125} \times 30}{\sqrt[3]{27}}$

B

4 Evaluate these.

a 5^{-1} b 6^0 c 3^{-2} d 4^1

e 18^0 f 8^{-1} g 11^1 h 6^{-2}

B A03

5 Sharlene has a piece of string.
She cuts the piece of string into quarters.
She then cuts one of the quarters into quarters.
She does this two more times.

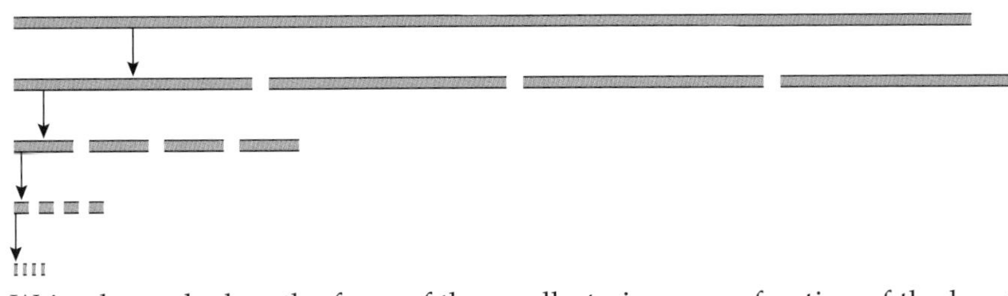

Write down the length of one of the smallest pieces as a fraction of the length of the first piece

a as a repeated multiplication

b as a power of $\frac{1}{4}$

c as a power of 4.

10.5 Prime factors

C

1 In each case work out the product of the prime factors.

a $2 \times 5 \times 7$ b $2 \times 2 \times 3 \times 5$

2 Work out the numbers represented by

a $2^2 \times 5$ b $2 \times 3^3 \times 7$

3 Use a factor tree to write each of these numbers as a product of its prime factors.

a 32 b 75

4 Write each of these numbers as a product of its prime factors.

a 84 b 140

5 a Write 42 as a product of its prime factors.

b Write 70 as a product of its prime factors.

c Write down all the common prime factors of 42 and 70.

d What is the HCF of 42 and 70?

6 Use prime factors to work out the HCF of 90 and 150.

7 a Write 30 as a product of its prime factors.
b Write 66 as a product of its prime factors.
c Write down all the common prime factors of 30 and 66.
d What is the LCM of 30 and 66?

C

8 Use prime factors to work out the LCM of 15 and 36.

9 Work out the HCF of 210 and 300.

C

10 Work out the LCM of 42 and 45.

AO2

11 Lily has some counters. She puts them into bags, 18 counters in each bag. There are none left over.
She starts again and puts them into bags, 15 counters in each bag. There are none left over.
What is the smallest possible number of counters she could have?

C

AO3

10.6 Laws of indices and standard form

C

1 Write each of these expressions as a single power.

a $3^4 \times 3^7$ b $5^2 \times 5^8$ c $4^3 \times 4^9 \times 4^2$

d $6^5 \div 6^2$ e $7^9 \div 7^5$ f $\frac{9^7}{9^3}$

B

2 Write each of these expressions as a single power. Then work out its value.

a $\frac{4^3 \times 4^2}{4^4}$ b $\frac{3^6 \times 3^3}{3^5}$ c $\frac{2^7}{16}$

3 Simplify and evaluate each of these expressions.

a $2^3 \times 4^5 \times 2^{-1} \times 4^{-3}$ b $5^5 \times 3^4 \times 3^{-2} \times 5^{-3}$

c $\frac{4^5 \times 2^3}{4^4 \times 2}$ d $\frac{6^6 \times 5^9}{6^5 \times 5^7}$

4 You are given that $1\,048\,576 = 2^4 \times 4^8$ and $8192 = 2^3 \times 4^5$.
Work out $1\,048\,576 \div 8192$ without a calculator.

5 Write each of these numbers in standard form.

a 452 000 b 0.006 7

c 2 400 500 d 0.000 000 329

6 Write each of these as a decimal number.

a 3.49×10^4 b 2.7×10^{-4} c 6×10^9 d 3.08×10^{-7}

7 Work out these. Give your answer in standard form.

a $(4 \times 10^3) \times (2 \times 10^7)$ b $(2 \times 10^5) + (6 \times 10^5)$

c $5 \times 10^{-2} \times 7 \times 10^5$ d $(2 \times 10^5) - (6 \times 10^4)$

e $(5 \times 10^4) \div (2 \times 10^8)$ f $(4 \times 10^{-5}) \div (8 \times 10^{-2})$

g $(5 \times 10^4) + (5 \times 10^5)$ h $(5 \times 10^5) - (5 \times 10^4)$

B

8 A scientist is working out energy consumption projections for the UK. In 2009 the population of the UK was estimated to be 6×10^7. The total UK energy consumption in 2009 was approximately 400 billion kilowatt hours. Approximately what was the average energy consumption, in kilowatt hours, per person in the UK in 2009?

A billion is 1000 million.

AO2

Basic rules of algebra

Links to:
Middle Student Book
Ch11, pp.179–195

Key Points

Collecting like terms E

This expression $3a + 6b + 5a$ has three terms: $3a$, $6b$ and $5a$.

Terms that use the same letter are called like terms.

You can simplify algebraic expressions by collecting like terms together.

$3a + 6b + 5a = 3a + 5a + 6b = 8a + 6b$

Multiplying terms E

To multiply algebraic terms, you multiply the numbers and then multiply the letters.

$3f \times 4g = 3 \times f \times 4 \times g = 3 \times 4 \times f \times g = 12fg$

Expanding brackets D

To expand brackets, you multiply each term inside the bracket by the term outside the bracket.

$3(2a + 7) = 6a + 21$ $\quad$ $x(3x + 5) = 3x^2 + 5x$

Factorising D

Factorising an algebraic expression is the opposite of expanding brackets.

Start by writing a common factor of both terms outside a bracket.

factorising

$8t - 10 = 2(4t - 5)$

expanding

Then work out the terms inside the bracket.

Expanding two brackets C B

To expand two brackets, you multiply each term in one bracket by each term in the other bracket.

Method 1 – the grid method

$(x + 2)(x + 5)$

$\times$	x	5
x	x^2	$5x$
2	$2x$	10

Then add the terms together and simplify.

$(x + 2)(x + 5) = x^2 + 5x + 2x + 10 = x^2 + 7x + 10$

Method 2 – FOIL (Firsts, Outers, Inners, Lasts)

$(x + 2)(x + 5)$

Firsts: $x \times x = x^2$

Outers: $x \times 5 = 5x$

Inners: $2 \times x = 2x$

Lasts: $2 \times 5 = 10$

Then add the terms together and simplify.

$(x + 2)(x + 5) = x^2 + 5x + 2x + 10 = x^2 + 7x + 10$

11.1 Collecting like terms

E

1 Simplify these expressions by collecting like terms.

a $2a + 3z + 4a$ $\quad$ b $6b + 6y + 6b$

c $3c + 4 + 5c + 6$ $\quad$ d $6d + 10x - 2d$

e $7e + 2w - 6e$ $\quad$ f $8f + 7 - 6f - 5$

2 Simplify these expressions.

a $2a + 4b + 6a + 8b$ $\quad$ b $8t + 3r - 2t + 7r$

c $4p + 7q - p - 5q$ $\quad$ d $8g - 4c - 3g - 2c$

e $3e + 4e + 5f + 6e + 7f$ $\quad$ f $3p - 2q + 7r - 4q + 2r - p$

3 Simplify these expressions.

a $2ab + 6a + 5ab - a$ $\quad$ b $3xy + 3y^2 - xy + 3y^2$

c $10t^2 - 9t + t^2$ $\quad$ d $5p^2 + 7 - 3p - 3$

e $4x^2 + xy - y^2 + x^2 + 4y^2$ $\quad$ f $3a^2 + 5b^2 - b^2 + 2ab + 4b^2$

4 Write an expression for the perimeter of each of these shapes.
Write each expression in its simplest form.

a a square, side $2x + 1$

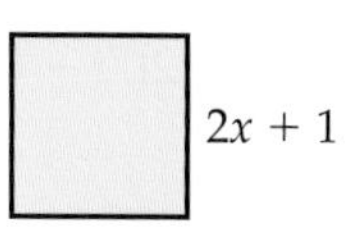

b a rectangle, width w, length $3t$

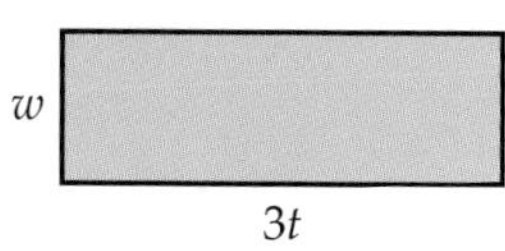

c a pentagon

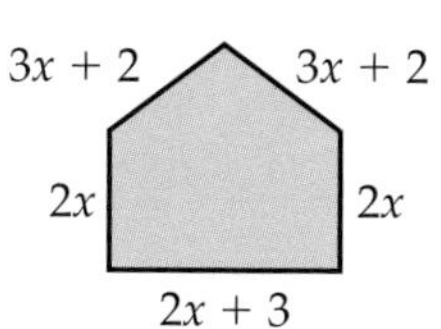

d an arrowhead

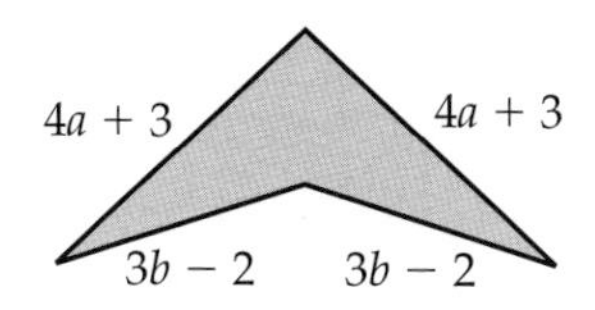

E

AO2

5 Complete this magic square.

All rows, columns and diagonals must add to the same total.

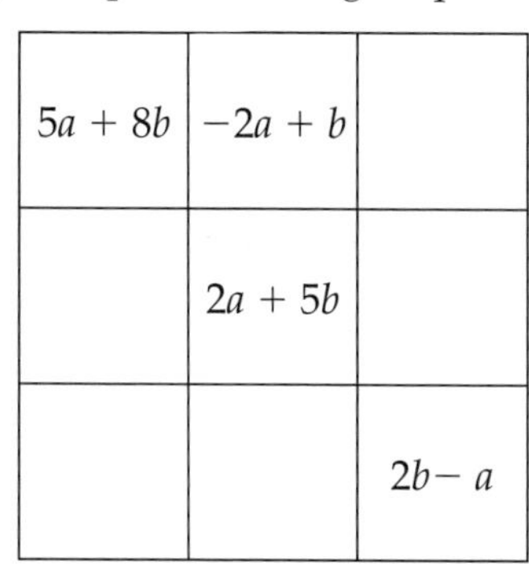

$5a + 8b$	$-2a + b$	
	$2a + 5b$	
		$2b - a$

E

AO3

11.2 Multiplying in algebra

1 Simplify the following expressions.

a $3 \times 2a$ b $5 \times 4b$ c $10 \times 6c$
d $3 \times d$ e $e \times 3$ f $3f \times 2$
g $5g \times 2x$ h $2h \times y$ i $i \times i$
j $3j \times j$ k $k \times 5k$ l $3l \times 4l$

E

11.3 Expanding brackets

1 Expand the following brackets.

a $3(a + 4)$ **b** $5(3 + b)$ **c** $6(c + x)$
d $4(d - 2)$ **e** $6(5 - e)$ **f** $9(f - y)$
g $5(3 + g + w)$ **h** $2(h + 2 - v)$ **i** $10(i - u - 4)$

2 Expand the following expressions.

a $2(3a + 6)$ **b** $3(4 + 2b)$ **c** $7(2c + x)$
d $6(4d - 2)$ **e** $5(5e - y)$ **f** $3(4f + 5w)$
g $4(3g + 2v - 3)$ **h** $6(2h - 3u + t)$ **i** $5(3i - 4j + k - 6)$
j $3(x^2 + 3x)$ **k** $4(x^2 - 4x + 1)$ **l** $5(2y^2 - y - 2)$

D

D

3 Expand these brackets.

a $-3(2a + 4)$ b $-4(2 + 5b)$ c $-5(2c - 4)$

d $-10(3 + 7d)$ e $-1(2e + 5)$ f $-2(7f - 8)$

4 Multiply out the brackets.

a $5(a^2 + 2a + 3)$ b $2(b^2 + 5b - 6)$ c $10(x^2 - 4x + 1)$

5 Expand the following brackets.

a $a(a + 2)$ b $b(b - 5)$ c $c(2c + 6)$

d $d(5d + z)$ e $e(2e - p)$ f $5f(f + 2)$

g $3g(3g - 3)$ h $2h(3t + 2r)$ i $5i(4r - 2q)$

D AO2

6 i Write an expression for the area of each shape using brackets.

ii Expand the brackets.

a

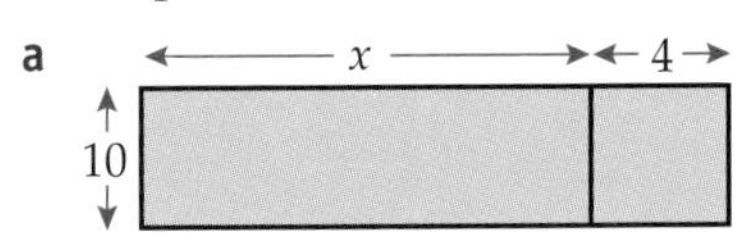

b

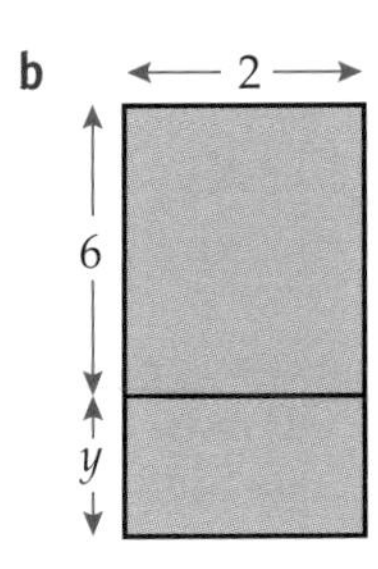

Area of rectangle = length × width

11.4 Simplifying expressions with brackets

D

1 Expand and simplify these expressions.

a $2(a + 5) + 4a - 3$ b $6(b - 2) - 3b + 15$

c $3(c + 2) + 4(c + 5)$ d $6(d + 2) + 10(d - 1)$

D AO2

2 Show that

a $3(a + 4) - a = 2(a + 6)$ b $4(b - 3) + 2 = 2(2b - 5)$

C

3 Expand and simplify these expressions.

a $3(2a + 4) + 2(a + 3)$ b $5(x - 1) + 3(3x - 6)$

c $5(2c + 3) - 4(2c + 3)$ d $3(d - 2) - 4(3d - 5)$

C AO2

4 Show that

a $8(a + 2) + 4(a + 1) = 4(3a + 5)$

b $5(3b + 1) - 3(b - 4) = 6(2b + 3) - 1$

11.5 Factorising algebraic expressions

D

1 Write down the common factor of each pair.

a $4x$ and 8 b $6x$ and 21 c p^2 and $5p$

2 Copy and complete. Check your answers by expanding the brackets.

a $3x + 9 = 3(\square + 3)$ b $4y - 12 = 4(\square - 3)$

c $7u + 21 = 7(u + \square)$ d $5h - 55 = \square(h - 11)$

e $6m - 24 = \square(m - 4)$ f $3t + 27 = \square(t + \square)$

3 Factorise these expressions.

a $5x + 20$ b $2y - 10$ c $4z + 4$

d $7a + 28$ e $4 + 2b$ f $45 - 9p$

D

4 Write whether A, B or C is the correct factorisation of each of these expressions.

a $x^2 + 2x =$ A $x(x + 2x)$ B $x(x + 2)$ C $x(2x + 2)$

b $f^2 - 10f =$ A $f(f - 10f)$ B $f^2(1 - 10f)$ C $f(f - 10)$

c $3y + y^2 =$ A $y(3 + y)$ B $3(y + y^2)$ C $y(3 + 2)$

d $7p^2 + p =$ A $7p(p + 1)$ B $7(p^2 + p)$ C $p(7p + 1)$

5 There are six numbers missing from the factorisations below.
There are 8 number cards on the right.

D

$4x + 6 = \square(2x + \square)$

$8y - 20 = \square(2y - \square)$

$14m^2 + 8m = 2m(\square m + 4)$

$18g - 2g^2 = 2g(\square - g)$

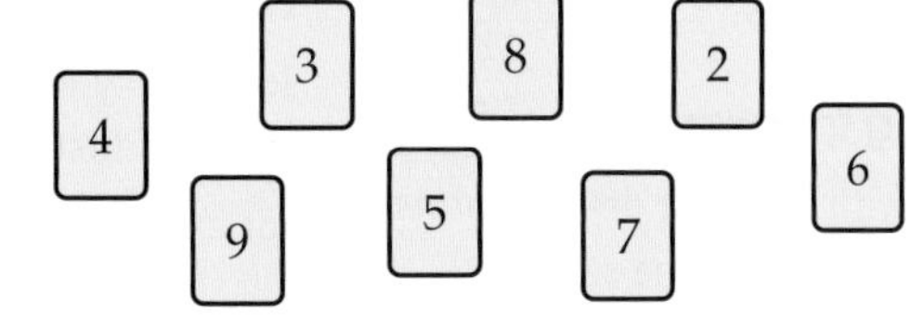

What is the product of the two number cards that are not used in the factorisations?

A02

11.6 Expanding two brackets

1 Expand and simplify these expressions.

C

a $(x + 2)(x + 3)$ b $(x + 3)(x - 2)$ c $(x - 1)(x + 4)$

d $(x + 6)(x - 6)$ e $(x - 5)(x + 5)$ f $(x + y)(x - y)$

g $(x + 3)^2$ h $(x - 4)^2$ i $(x - a)^2$

2 Here is part of a number grid.

C

1	2	3	4	5
6	7	8	9	10
11	12	13	14	15
16	17	18	19	20

Maria is investigating blocks of six numbers.
Here is a block of six numbers, with n in the left-hand corner.

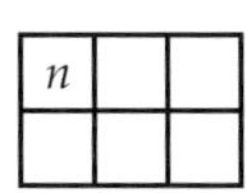

a Copy this block of six numbers and write expressions in terms of n in the other five boxes.

b Multiply together the term in the bottom left box of your block of six and the term in the top right.

c Multiply together the term in the top left box of your block of six (i.e. n) and the term in the bottom right.

d Subtract your answer to part c from your answer to part b. What do you notice?

A02

3 Expand and simplify these expressions.

B

a $(5a + 3)(2a + 4)$ b $(3b + 6)(4b - 2)$ c $(3c - 5)(4c - 4)$

d $(4d - 1)(4d + 1)$ e $(2e + 3)(2e - 3)$ f $(4 - 2f)(4 + 2f)$

Links to:
Middle Student Book
Ch12, pp.198–211

Key Points

Comparing fractions — E D

To compare fractions with different denominators, change them into equivalent fractions with the same denominator.

Multiplying fractions — E

To multiply fractions, multiply the numerators together and multiply the denominators together.

Before you multiply, cancel common factors if possible.

For example, $\frac{\cancel{4}^{2}}{{}_{1}\cancel{5}} \times \frac{\cancel{15}^{3}}{\cancel{22}_{11}} = \frac{2}{1} \times \frac{3}{11} = \frac{2 \times 3}{1 \times 11} = \frac{6}{11}$

Adding and subtracting fractions — E D

To add or subtract fractions with different denominators, write them as equivalent fractions with the same denominator, then add or subtract the numerators.

For example, $\frac{1}{2} + \frac{1}{3} = \frac{3}{6} + \frac{2}{6} = \frac{5}{6}$

Adding and subtracting mixed numbers — C

You can add or subtract mixed numbers by changing them into improper fractions.

For example, $1\frac{1}{3} + 3\frac{2}{5} = \frac{4}{3} + \frac{17}{5} = \frac{20}{15} + \frac{51}{15} = \frac{71}{15} = 4\frac{11}{15}$

Finding reciprocals — C

When two numbers can be multiplied together to give the answer 1, then each number is called the reciprocal of the other.

The reciprocal of a fraction is found by turning the fraction upside down.

The reciprocal of a number is 1 divided by that number.

Multiplying fractions and mixed numbers

To multiply mixed numbers, change them to improper fractions first.

For example, $4\frac{1}{3} \times \frac{2}{5} = \frac{13}{3} \times \frac{2}{5} = \frac{13 \times 2}{3 \times 5} = \frac{26}{15} = 1\frac{11}{15}$

Dividing by a fraction — D C B

To divide by a fraction, turn the fraction upside down and multiply.

When the division involves mixed numbers, change them to improper fractions first.

For example, $\frac{3}{4} \div 7 = \frac{3}{4} \div \frac{7}{1} = \frac{3}{4} \times \frac{1}{7} = \frac{3 \times 1}{4 \times 7} = \frac{3}{28}$

12.1 Comparing fractions

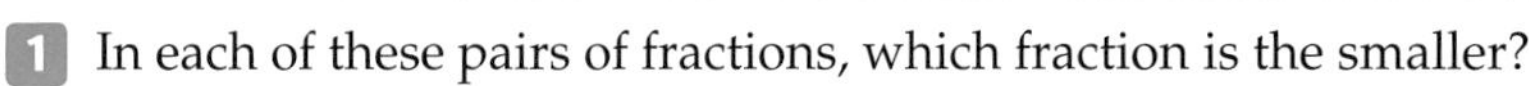

E

1 In each of these pairs of fractions, which fraction is the smaller?

a $\frac{4}{5}, \frac{3}{10}$ b $\frac{7}{11}, \frac{5}{8}$

2 Which of these fractions is closer to $\frac{1}{3}$?

A $\frac{5}{12}$ **B** $\frac{3}{8}$

Give a reason for your answer.

E AO2

3 Shona says, '$\frac{5}{6}$ is bigger than $\frac{22}{30}$.'

Is Shona correct? Give a reason for your answer.

D AO2

4 Sam has three fraction cards.

He puts them in order of size, starting with the smallest.

This is the order he puts them in.

Is Sam correct? Give a reason for your answer.

$\frac{7}{12}$ $\frac{9}{16}$ $\frac{15}{24}$

D AO3

5 Ted says:

I am thinking of a fraction. My fraction is bigger than $\frac{3}{5}$ but smaller than $\frac{4}{5}$. The numerator and denominator of my fraction are both two-digit odd square numbers. What is the fraction I am thinking of?

1 Work out

a $\frac{4}{5} - \frac{3}{10}$ b $\frac{2}{3} - \frac{1}{6}$ c $\frac{3}{4} - \frac{5}{12}$ d $\frac{2}{3} + \frac{2}{9}$

2 Mo has a bag of rice. She uses $\frac{1}{4}$ kg of the rice in a risotto.
She now has only $\frac{1}{8}$ kg of rice left in the bag.
How much rice did Mo start with in the bag?

3 Jane decorates cakes.
She uses $\frac{4}{5}$ m of ribbon to go around a cake and $\frac{3}{20}$ m of ribbon for the bow.

a What is the total length of ribbon Jane uses for one cake?

b Jane has an order for 5 cakes. Cake ribbon costs 72p per metre. What is the total cost of the ribbon for the 5 cakes?

4 Andy has three fractions cards.

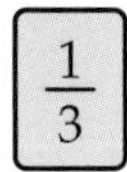

All the fractions are positive.
Two of the fractions are $\frac{1}{3}$ and $\frac{5}{12}$.
Andy adds the three fractions and gets an answer of $\frac{11}{12}$.
Could Andy's answer be correct? Show working to support your decision.

5 Work out

a $\frac{2}{3} - \frac{1}{4}$ b $\frac{1}{5} + \frac{2}{7}$ c $\frac{8}{9} - \frac{5}{12}$ d $\frac{3}{8} + \frac{2}{5}$

6 Greg wants to make a shelf that is 1 m long.
He has two pieces of wood. One is $\frac{3}{5}$ m long and the other is $\frac{3}{8}$ m long.
If Greg joins the two pieces of wood together, will the total length be long enough to make the shelf? Show working to support your decision.

7 Ali adds together two proper fractions.
Each fraction has a different denominator.
He gets an answer of $1\frac{2}{15}$.
Suggest two fractions that Ali may be using.

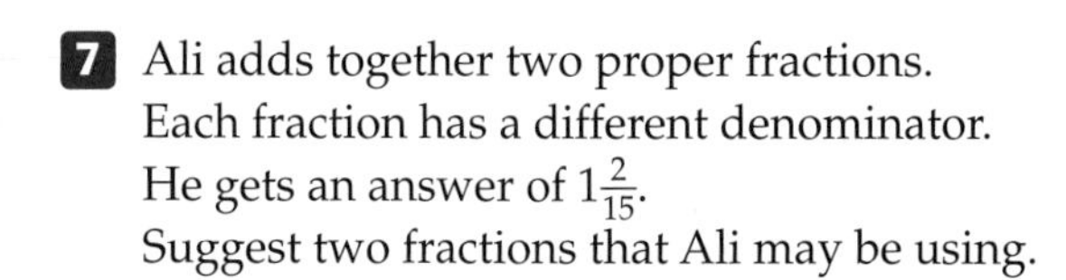

8 Chad has two fraction cards.

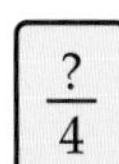

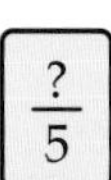

Both the fractions are positive.
Chad adds the two fractions and gets an answer that cancels to $\frac{2}{5}$.
Chad is wrong. Show working to explain why.

9 Tammy has three fractions cards.

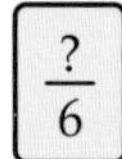

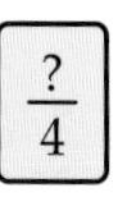

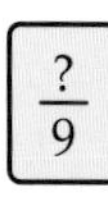

All the fractions are positive.
Tammy adds the three fractions and gets an answer that cancels to $\frac{1}{2}$.
Could Tammy's answer be correct? Show working to support your decision.

12.3 Adding and subtracting mixed numbers

C

1 Work out these.

a $4\frac{1}{2} + 2\frac{3}{8}$

b $3\frac{4}{5} + 1\frac{3}{10}$

c $2\frac{1}{3} + 5\frac{3}{5}$

d $5\frac{1}{2} - 2\frac{3}{10}$

e $2\frac{4}{5} - 2\frac{3}{4}$

f $5\frac{4}{7} - 1\frac{5}{6}$

C AO2

2 Anna is training for a marathon.
The distance from her work to her home is $12\frac{1}{4}$ miles.
Each evening she leaves work and walks the first $1\frac{3}{5}$ miles home.
She runs the rest of the way. How far does she run?

C AO3

3 Sarifa has two mixed-number cards.

$2\frac{?}{8}$

Both the mixed numbers are positive.
Sarifa adds the mixed numbers and gets an answer that cancels to $6\frac{1}{4}$.
Could Sarifa's answer be correct?
Show working to support your decision.

C AO2

4 Sandra wants to buy a horse trailer. She sees one advertised in the local paper. The trailer can carry two horses.
The maximum weight that her car can pull is $2\frac{3}{5}$ tonnes.
The trailer weighs $\frac{9}{10}$ tonne. Sandra's horses weigh $\frac{1}{2}$ tonne and $\frac{4}{5}$ tonne.
Is this trailer suitable for Sandra? Show working to support your decision.

12.4 Multiplying fractions

E

1 Work out these multiplcations.

a $\frac{1}{4} \times \frac{1}{3}$

b $\frac{2}{5} \times \frac{1}{8}$

c $\frac{3}{7} \times \frac{2}{9}$

d $\frac{4}{5} \times \frac{15}{16}$

2 a Work out $\frac{5}{7} \times \frac{2}{5}$ and $\frac{3}{8} \times \frac{5}{6}$. Write each answer in its simplest form.

b Write your answers to part a as equivalent fractions with the denominator 112.

c Which is the larger fraction, $\frac{5}{7} \times \frac{2}{5}$ or $\frac{3}{8} \times \frac{5}{6}$?

E AO2

3 Which is the smaller fraction, $\frac{4}{5} \times \frac{3}{7}$ or $\frac{3}{5} \times \frac{7}{9}$?

E AO3

4 In class 10T, none of the boys wear glasses.
$\frac{1}{2}$ of the students in 10T are girls.
$\frac{2}{3}$ of the girls are right-handed.
$\frac{1}{5}$ of the right-handed girls wear glasses.
None of the left-handed girls wear glasses.
What fraction of the whole class does not wear glasses?

12.5 Multiplying mixed numbers

1 Work out these multiplications.

Give your answers as mixed numbers when possible.

a $4 \times 2\frac{3}{5}$ **b** $3 \times 1\frac{7}{8}$ **c** $6\frac{1}{4} \times 2$ **d** $3\frac{4}{7} \times 3$

2 Work out these multiplications.

Simplify your answers, and write them as mixed numbers when possible.

a $\frac{1}{3} \times 2\frac{2}{5}$ **b** $\frac{3}{5} \times 3\frac{2}{3}$ **c** $4\frac{1}{4} \times \frac{4}{5}$ **d** $2\frac{2}{3} \times \frac{5}{8}$

3 Rhys is going to fit some laminate flooring.
Each strip of flooring is $\frac{3}{8}$ m wide.
The length of the flooring Rhys needs $43\frac{1}{2}$ m.

a What is the area of the laminate flooring that Rhys needs?

> **Area of rectangle = length × width**

The laminate flooring costs £37.50 per square metre.
It can only be bought in a whole number of square metres.

b How much does Rhys pay for the laminate flooring?

4 Sammy is a home worker.
She packs information leaflets into envelopes.
Sammy packs, on average, a dozen envelopes in $3\frac{1}{2}$ minutes.
Sammy must pack 1 080 envelopes in a day to get a full day's pay.
Approximately how long will it take Sammy to earn a full day's pay?

> **First find $\frac{1}{12}$ of $3\frac{1}{2}$ minutes to find out how long it takes Sammy to pack one envelope.**

5 Work out these multiplications.

Simplify your answers, and write them as mixed numbers when possible.

a $2\frac{2}{3} \times 1\frac{3}{4}$ **b** $1\frac{4}{5} \times 4\frac{2}{3}$ **c** $4\frac{1}{6} \times 2\frac{1}{5}$ **d** $2\frac{2}{7} \times 3\frac{3}{8}$

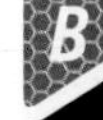

6 A lawn measures $5\frac{1}{4}$ m by $6\frac{2}{3}$ m.
What is the area of the lawn?

12.6 Reciprocals

1 **a** Find the reciprocal of each of these numbers.
Give your answers as fractions, whole numbers or mixed numbers.

i 6 **ii** 12 **iii** 50 **iv** $\frac{1}{3}$ **v** $\frac{2}{7}$ **vi** $\frac{4}{5}$

b Check your answers to part **a** by multiplying each number by its reciprocal. The answer to each multiplication should be 1.

2 Find the reciprocal of each of these numbers.

Give your answers as fractions, whole numbers or mixed numbers. Show all your working.

a 0.4 **b** 0.9 **c** 0.05 **d** 4.5

3 This is part of Sanisha's homework.
Has Sanisha got this question right or wrong?
Explain your answer.

> Q1 Find the reciprocal of 0.032
>
> $0.032 = \frac{32}{100} = \frac{32 \div 4}{100 \div 4} = \frac{8}{25}$
>
> Reciprocal $= \frac{25}{8} = 3\frac{1}{8}$

D

1 Work out these divisions.

a $12 \div \frac{1}{2}$ b $6 \div \frac{1}{3}$ c $\frac{2}{5} \div 2$ d $\frac{3}{8} \div 4$

2 Work out these divisions.

a $\frac{3}{4} \div \frac{2}{3}$ b $\frac{4}{5} \div \frac{3}{7}$ c $\frac{8}{9} \div \frac{2}{3}$ d $\frac{12}{17} \div \frac{6}{7}$

D

3 Harry shares $2\frac{1}{2}$ apple pies between 6 people.
How much do they each receive?

4 Penny has a recipe for cheesecake.
One cheesecake needs $\frac{1}{3}$ pint of cream.

Penny has $3\frac{1}{4}$ pints of cream in her fridge.

a What is the greatest number of cheesecakes that Penny can make with the cream in her fridge?

b Penny makes the number of cheesecakes found in part **a**. Altogether how much cream does she use?

AO2

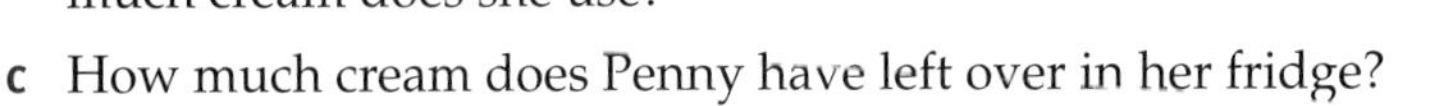

c How much cream does Penny have left over in her fridge?

FUNCTIONAL

C

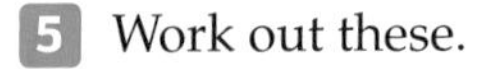

5 Work out these.

a $1\frac{1}{2} \div \frac{3}{5}$ b $5\frac{1}{4} \div 2\frac{2}{3}$ c $1\frac{1}{3} \div \frac{24}{13}$ d $\frac{1}{4} \div 1\frac{3}{8}$

C

6 Jane shares $2\frac{1}{4}$ chocolate cakes between 6 people.
How much do they each receive?

AO2

7 Paul pours $3\frac{1}{2}$ litres of peach smoothie into $\frac{1}{5}$ litre glasses.
How many full glasses does he pour?

B

8 a Work out these divisions.

i $6\frac{2}{3} \div 1\frac{4}{5}$ ii $3\frac{5}{6} \div 2\frac{1}{4}$

Write your answers in their simplest form.

AO2

b Use your answers to part **a** to work out $6\frac{2}{3} \div 1\frac{4}{5} - 3\frac{5}{6} \div 2\frac{1}{4}$

Remember BIDMAS.

B

9 Meg says, 'If I divide $3\frac{1}{4}$ by $1\frac{5}{6}$, my answer will be less than $1\frac{5}{6}$.'
Is Meg correct? Show working to support your answer.

AO3

Links to:
Middle Student Book
Ch13, pp.212–222

Key Points

Adding and subtracting decimals E

To add or subtract decimals, set out the calculation in columns by lining up the decimal points, then add or subtract.

Converting decimals to fractions D

You can use place value to convert a decimal to a fraction.

First write the decimal as tenths, hundredths, thousandths, …

Take the biggest denominator and put it on the bottom of the fraction. Put the decimal digits on the top of the fraction. Remember to give your answer in its simplest form.

Terminating decimals D

Terminating decimals are decimals that come to an end, for example 0.6.

Multiplying decimals D

To multiply decimals

1 ignore the decimal points and multiply the numbers
2 count the decimal places in the calculation
3 put this number of decimal places in the answer.

Dividing decimals D C

To divide a decimal by a whole number, line up the decimal point in the answer and divide in the normal way.

To divide by a decimal, first write the division as a fraction. Convert the denominator to a whole number and create an equivalent fraction. Then divide in the usual way.

Recurring decimals C B

Recurring decimals never end, for example 0.333…

'Recurring dots' are used to make the pattern clear. A dot over one digit shows that this digit recurs. Two dots, at the beginning and end of a sequence of digits, show that the sequence recurs.

13.1 Adding and subtracting decimals

1 A plumber has some pieces of copper piping with lengths 4 m, 3.25 m, 1.1 m and 0.35 m.
What is the total length of the copper piping?

E

2 Andrea went to a restaurant for a meal. She had a starter for £3.95, a main course for £13.50, a pudding for £6 and a drink for £0.99.

a What was the total cost of her meal?

b Andrea paid with a £20 note and a £10 note.
How much change did she receive?

E

3 A carpenter has a 5 m length of skirting board.
He cuts off three pieces. The pieces are 1.75 m, 1.35 m and 1.5 m long.
What length of skirting board does he have left?

4 Which two of these numbers are closest together in value?

7.24 | 10.04 | 17 | 21.6

5 George made four deliveries for his company. This is his timesheet.

From:		To:	Miles:
Office	⟶	Mrs Grayson	26.4
Mrs Grayson	⟶	Mr Black	17.2
Mr Black	⟶	Miss Carlett	31.6
Miss Carlett	⟶	Office	21.1

Car mileage: Start 021336.8 Finish ________

Work out what the finish mileage should be. Show all your workings.

AO2

13.2 Converting decimals to fractions

D

1 Convert each decimal to a fraction. Give the fraction in its lowest terms.

a 0.3 **b** 0.2 **c** 0.02 **d** 0.15

e 0.125 **f** 0.003 **g** 0.055 **h** 0.0004

2 Convert these to mixed numbers.

a 7.5 **b** 7.8 **c** 7.32 **d** 7.95

3 Last year 0.65 of all students had a grade C or above in their GCSE maths.
What fraction of the students had a grade C or above?

4 Sofia did a survey in her maths class.
She found that 0.6875 of her classmates had the recommended calculator.
What fraction of her classmates had the recommended calculator?

13.3 Multiplying and dividing decimals

D

1 Dennis buys 1.4 kg of grapes, costing £2.35 per kg.
Work out the cost of the grapes.

2 Evelyn has to buy 17 pencils. They each cost £0.52.
What is the total cost of the pencils?

3 Work out these.

a 6.24×0.05 **b** 62.4×0.005

AO2

c What do you notice about your answers to parts **a** and **b**?

C

4 Work out these.

a $9 \div 0.03$ **b** $25 \div 0.005$ **c** $25.926 \div 0.6$

C

5 Mrs Jones is putting paving slabs on the ground along one side of her greenhouse, as shown in the diagram.
Each paving slab is 0.6 m long.
The greenhouse is 10.8 m long.
The paving stones come in packs of four.
How many packs must she buy?

AO2

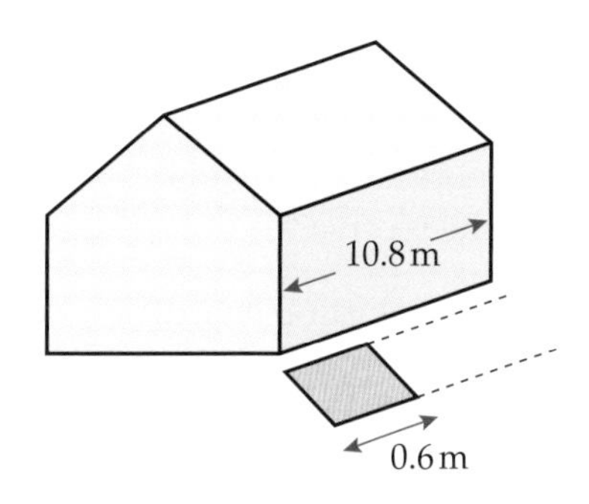

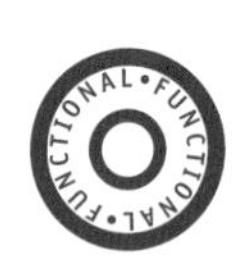

13.4 Converting fractions to decimals

D

1 Write these fractions and decimals in order of size, smallest first.

$\frac{7}{10}$ 0.82 $\frac{3}{4}$ 0.599 $\frac{5}{8}$ 0.622

> **Start by converting all the fractions into terminating decimals.**

C

2 Write these fractions in order of size, smallest first.

$\frac{2}{3}$ $\frac{7}{9}$ $\frac{5}{6}$ $\frac{7}{11}$ $\frac{5}{7}$

> **Start by converting all the fractions into recurring decimals.**

B

3 Convert each of these fractions to a decimal.

a $\frac{5}{13}$ **b** $\frac{6}{13}$ **c** $\frac{7}{13}$ **d** $\frac{8}{13}$

AO2

What do you notice?

Equations and inequalities

Links to:
Middle Student Book
Ch14, pp.223–241

Key Points

Solving-two step equations

E D

You will need to use two or more steps to solve some equations.

For example, to solve $2x + 5 = 11$:

$$2x + 5 - 5 = 11 - 5$$
$$2x = 6$$
$$2x \div 2 = 6 \div 2$$
$$x = 3$$

Writing and solving equations

E D

When writing equations decide what your unknown will represent and then write an expression. Turn this into an equation by looking for expressions or numbers that are equal.

Equations with brackets

When you solve equations with brackets expand all the brackets first.

Equations with an unknown on both sides

To solve this type of equation you need to get the unknown on one side of the equals sign only.

For example, $3x - 3 = 2x + 5$.

$$3x - 2x - 3 = 2x - 2x + 5$$
$$x - 3 = 5$$

Then solve the equation as normal.

Inequalities

E D C B

$x > 2$ is an inequality. It means that x must be greater than 2. It can be shown on a number line.

0 1 2 3 4 5 6 7

The open circle means that the 2 is not included.

$x \leqslant 5$ means that x is less than or equal to 5. The closed circle shows that in this case, the 5 is included in the set of solutions.

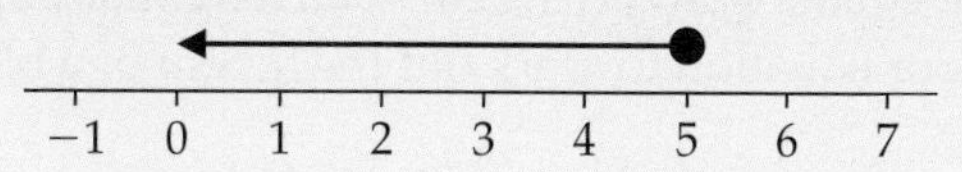

You can use the balance method to solve inequalities.

When you multiply or divide both sides of an inequality by a positive number, the inequality sign remains unchanged.

When you multiply or divide both sides of an inequality by a negative number, you need to change the direction of the sign.

Simultaneous equations

B

To solve simultaneous equations you eliminate one of the unknowns using algebraic steps. You should then have a linear equation with one unknown, which you can solve.

14.1 Solving two-step equations

E

1 Solve these equations.

a $3a + 6 = 18$ b $5b + 3 = 28$ c $4c - 3 = 13$

d $2d - 5 = 11$ e $2 + 4e = 22$ f $2f - 3 = -9$

D

2 Solve the following.

a $\frac{a}{5} + 2 = 5$ b $\frac{b}{2} - 6 = 3$ c $5 + \frac{c}{3} = 9$

3 Solve these equations. Give your answers as mixed numbers.

a $3a = 13$ b $3b - 4 = 13$ c $13 = 5c + 4$

4 Solve these equations. Give your answers as decimal numbers.

a $2a + 3 = 4$ b $4b - 5 = 6$ c $8 + 9c = 7$

d $18d - 8 = 2$ e $12 = 4e + 11$ f $13 = 6 + 5f$

E

1 I think of a number and divide it by 6. My answer is 5.

a Use algebra to make an equation.

b Solve your equation to find the number I was thinking of.

AO2

2 I think of a number and multiply it by 4. My answer is 80.
Use algebra to work out the number I was thinking of.

D

3 Fern builds this tower from blue and yellow blocks.
A yellow block is 3 cm high, and a blue block is x cm high.
The total height of her tower is 19 cm.
Write an equation involving the total height of the tower.
Solve your equation to work out the height of a blue block.

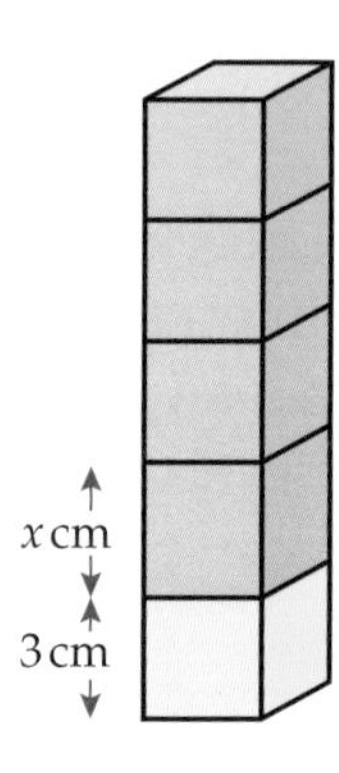

4 a Write an expression for the perimeter of this pentagon.
All lengths are in centimetres.

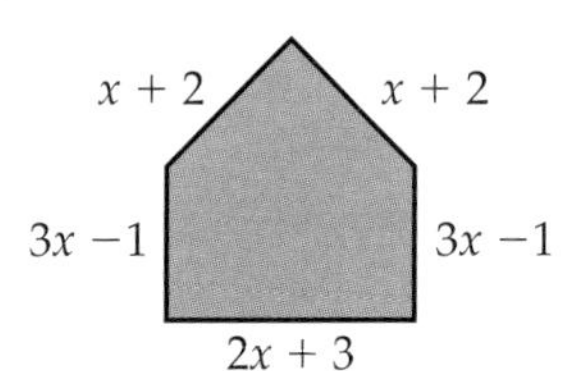

The perimeter is the distance around the outside of a shape.

b Simplify your expression in part **a**.

c The perimeter of the pentagon is 75 cm.
Write an equation and solve it to find the value of x.

5 a Explain why the sum of five consecutive numbers can be written as

$$n + n + 1 + n + 2 + n + 3 + n + 4$$

b Simplify this expression.

c The sum of five consecutive numbers is 105.
Use part **b** to write an equation, and solve it to find the value of the first number.

AO2

D

6 Billy is four times as old as his brother Joe.
The sum of their ages is 20 years.
Using x for Joe's age, use algebra to find out how old Joe is.

7 Carlos is going to buy a running machine. The total cost of the running machine is £845.
Carlos already has 20% of the money.
He knows he can save £35 per week.
Let x be the number of weeks that Carlos must save £35.
Write an equation and solve it to find the number of complete weeks it will take Carlos to save enough money to buy the running machine.

AO3

14.3 Equations with brackets

1 Solve the following equations.

a $2(z + 3) = 12$　b $3(y + 2) = 12$　c $4(w - 3) = 4$
d $3(v - 7) = -3$　e $4(2u + 3) = 28$　f $3(3t - 4) = -30$
g $51 = 3(5s + 2)$　h $12 = 4(2r - 3)$　i $-25 = 5(3q - 2)$

D

2 Solve these equations.

a $3(a + 4) + 3 = 21$　b $6(b + 2) - 8 = 28$
c $4(c - 2) + 6 = 10$　d $2(x + 3) + 5x = 20$
e $5(2x + 3) + 2x = 51$　f $4(3x - 6) - 5x = 11$
g $3(3j + 4) + 2 + j = 114$　h $3(2x - 7) - 2 - 3x = -53$

C

14.4 Equations with an unknown on both sides

1 Solve the following equations.

a $6x + 5 = 4x + 11$　b $9x - 13 = 6x + 2$　c $4x - 15 = 3x - 6$
d $6x + 9 = 3x + 24$　e $4x - 3 = x + 6$　f $8x - 5 = 3x + 10$
g $3x + 4 = 24 - 2x$　h $x + 4 = 12 - 3x$　i $9 - 2x = 2x + 1$

D

2 Solve these equations.

a $3a + 18 = 3(2a + 3)$
b $2(2b + 10) = 12(b - 1)$
c $3(6c - 2) = 4(10c - 7)$

C

3 This rectangle has an area of 189 cm².
Use algebra to find the length and the width of the rectangle.

$(8x - 3)$ cm
$(2x + 15)$ cm

C
AO3

14.5 Equations with fractions

1 Solve the following equations by first eliminating the denominator.

a $\frac{2a + 3}{3} = 3$　b $\frac{5b + 1}{7} = 3$　c $\frac{4c - 2}{2} = 5$
d $\frac{7d + 1}{10} = 5$　e $3 = \frac{10 - e}{2}$　f $3 = 9 - \frac{f}{4}$

C

2 Solve these equations.

a $\frac{x}{4} + \frac{3x}{4} = 3$　b $\frac{2x}{3} - \frac{x}{6} = 15$
c $\frac{x}{2} + \frac{3}{8} = \frac{19}{8}$　d $\frac{3x + 2}{4} = \frac{2x + 4}{3}$

B

3 Half of a number added to one fifth of the number is equal to six less than the number.
Use algebra to find out what the original number is.

B
AO3

14.6 Inequalities

E

1 Show each inequality on a number line.

a $x < 4$ b $x > 4$ c $x \leqslant 4$

d $x \geqslant -2$ e $-2 < x \leqslant 2$ f $x < 3$ or $x \geqslant 6$

Check that you have filled in the circle for part c.

D

2 x is an integer. List all the values of x such that:

a $3 < x < 6$ b $3 \leqslant x < 6$ c $-3 < x \leqslant 0$

d $-2 < x < 2$ e $-22 \leqslant x \leqslant -20$ f $-5 \leqslant x < -4$

C

3 Solve each of the following inequalities.

a $7x > 28$ b $3x + 2 < 11$ c $\frac{x}{4} \geqslant 3$

d $\frac{x}{3} + 2 \geqslant 6$ e $8x < 6x + 4$ f $x \leqslant 3x + 8$

4 Solve the inequality. Show the solution on a number line.

$6 < 3x \leqslant 18$

B

5 Solve the following inequalities.

a $\frac{x + 3}{4} \geqslant 2$ b $13x + 5 \leqslant 11x - 3$ c $-7 \leqslant 4x + 1 < 13$

6 Solve the following inequalities.

a $2 \leqslant 5x + 1 \leqslant 11$ b $-10 < 4x + 2 \leqslant 6$ c $0 \leqslant 2(x + 7) < 40$

14.7 Simultaneous equations

B

1 Solve the following pairs of simultaneous equations.

a $2x + y = 21$
$-2x + 2y = -12$

b $x + y = 6$
$3x - y = 10$

c $5x + 2y = 23$
$3x + y = 13$

d $3x + 4y = 10$
$2x - y = 3$

e $-3x - 2y = 0$
$4x + 4y = 4$

f $4x + 3y = 15$
$5x - 4y = 11$

g $2x + 7y = 17$
$3x + 5y = 20$

h $3x - 2y = 1$
$5x + 3y = -30$

B

2 Xavier is older than Yoris.
The sum of Xavier's and Yoris's age is 24 years. The difference in their ages is 6 years.

a Write equations for the sum of their ages and the difference in their ages.

b Solve your equations simultaneously.

c How old are Xavier and Yoris?

3 A bag contains 40 coins. All the coins are either 2p or 5p coins.
The value of the coins is £1.55.
Let x be the number of 2p coins, and y be the number of 5p coins.

a Write two equations involving x and y.

b Solve your equations simultaneously.

A02

c How many 2p coins are in the bag?

Indices and formulae

Links to:
Middle Student Book
Ch15, pp.242–258

Key Points

Writing Formulae E D

A formula is a general rule that shows the relationship between quantities. These quantities are called variables.

You can use letters for the variables in a formula.

For example $p = hr + b$
where p is the pay, h is the hours worked, r is the rate of pay and b is the bonus.

Substitution E D C

Use the correct order of operations to help you do the calculations when you substitute values into an algebraic expression.

Indices E D C

x^2 is called 'x squared', y^3 is called 'y cubed'.

$t \times t \times t \times t = t^4$. You say '$t$ to the power 4'.

The ^{4} is called the index.

Law of indices C

To **multiply** powers of the **same** number or variable **add** the indices.

$x^n \times x^m = x^{n+m}$

To **divide** powers of the **same** number or variable **subtract** the indices.

$x^n \div x^m = x^{n-m}$

Powers of 1 and 0 C B

Any number or variable raised to the power 1 is equal to the number or variable itself.

For example: $3^1 = 3$, $x^1 = x$

Any number or variable raised to the power 0 is equal to 1

For example: $3^0 = 1$, $x^0 = 1$

Changing the subject of a formula C B

The subject of a formula only appears once, and only on its own side of the formula.

In the formula $v = u + at$ the variable v is the subject.

You can rearrange a formula to make a different variable the subject. For example, you can rearrange $v = u + at$ as $a = \frac{v - u}{t}$

A power to a power B

To raise a power of a number or variable to a further power, multiply the indices.

$(x^n)^m = x^{nm}$

15.1 Using index notation

E

1 Simplify the following expressions using index notation.

a $x \times x \times x \times x$ b $y \times y$ c $c \times c \times c \times c \times c \times c$

D

2 Simplify these expressions using index notation.

a $7 \times x \times x \times x$ b $y \times y \times 8 \times y$ c $3 \times z \times 2 \times z \times 1 \times z$

3 Simplify these expressions using index notation.

a $5a \times 4a$ b $6x \times 2x \times x$ c $c \times 2c \times 2c \times 2c \times c$

C

4 Simplify the following expressions.

a $a^2 \times a^2$ b $b^3 \times b^5$ c $c^{30} \times c^{12}$

5 Simplify these expressions.

a $5x^2 \times 4x^4$ b $7b^2 \times b^3$ c $4c^4 \times 4c^4$.

6 Simplify these expressions.

a $x^5 \div x^3$ b $x^7 \div x^6$ c $c^{30} \div c^{12}$

7 Simplify these expressions.

a $12x^3 \div x^2$ b $16b^5 \div 8b^2$ c $20x^5 \div 5x$

C

8 Explain how you worked out your answer to Q4 part **c**.

AO3

9 Explain how you worked out your answer to Q6 part **c**.

15.2 Laws of indices (index laws)

C

1 Simplify each of the following expressions.

a $t^3 \times t^3$ b $t^4 \times t^4$ c $t^8 \div t^4$

d $t^3 \div t$ e $4t^6 \times t$ f $3t \times 3t^3$

g $5t^5 \div t$ h $12t^6 \div 6t^5$ i $6t^3 \div 6t^3$

2 Simplify each of these expressions.

a $\dfrac{a^3 \times a}{a^2}$ b $\dfrac{a^4 \times a^3}{a \times a^2}$ c $\dfrac{3a^2 \times 4a^3}{2a^4 \times a}$

B

3 Simplify each of these expressions.

a $5x^2y^2 \times 3xy^2$ b $2xy^3 \times 2xy^3$ c $3x \times 2x^2y^2 \times 2y^2$

4 Simplify each of these expressions.

a $p^2q^3 \div p$ b $10p^3q^3 \div 2p^2$ c $8p^4q^2 \div 8q^2$

d $\dfrac{9p^2q^4}{3p^2q^2}$ e $\dfrac{5p^4q^4}{5p^2q^4}$ f $\dfrac{10p^3q^4}{2p^3q}$

B

5 Simplify

a $(x^{12})^2$ b $(x^8)^3$ c $(x^4)^6$

AO2

d What do you notice about your answers to parts **a**, **b** and **c**?
Explain why this has happened.

15.3 Writing your own formulae

E

1 1Gb memory sticks cost £6 each.
Write a formula for the cost P, in pounds, for x 1Gb memory sticks.

2 George buys b bottles of water.
Each bottle of water costs 69p.
Write a formula for the total cost, C pence.

D

3 Pens cost p pence and pencils cost e pence.
James buys 12 pens and 5 pencils.
Write a formula for the total cost, T pence.

4 Isleta is training for a triathlon. She works out the time it takes her to do a training swim.

a For a medium training swim she allows $\frac{1}{2}$ a minute per length, plus an extra 5 minutes.
Write a formula for the time taken, in minutes, for Isleta to swim x lengths.

b For a hard training swim she allows 4 minutes plus $\frac{1}{3}$ of a minute per length.
Write a formula for the time taken, in minutes, for Isleta to swim x lengths.

5 An isosceles triangle has a base length of $4x$ and a perpendicular height of $5x$. Write a formula for the area, A, of this triangle.

D

15.4 Substituting into expressions

For the questions in this exercise use these values.

$a = 3, b = -2, c = 10$ and $d = \frac{1}{2}$

1 Work out the value of these expressions.

E

a $ac + d$ b $ab + c$ c $cd + ab$

d $3a + 6d$ e $5c + 3b$ f $3c - 4a + 5d$

g $\frac{a + 7}{5}$ h $\frac{c + b}{2}$ i $5(c - 2a)$

2 Work out the value of these expressions.

D

a $3a^2 - 2c$ **b** $b^2 + d$ **c** $3c^3 + d$

d $8d + c^2$ **e** $b^3 + c$ **f** $c(a + 2)$

g $\frac{3b + c}{2}$ **h** $\frac{c^2 + 10b}{10}$ **i** $\frac{2c + a^2 + 1}{5}$

3 Work out the value of these expressions.

C

a $c^2(2a + 2d)$ b $b(c^2 - a^2)$ c $c(a^2 - b^2)$

d $\frac{2a + c}{2a - b}$ e $\frac{2b + c}{8ad}$ f $\frac{2d + 5b}{a^2}$

g $\sqrt{c + 2a}$ h $\frac{\sqrt{4b^2 + a^2}}{c}$ i $\frac{\sqrt{ac - b^2 - 1}}{b}$

15.5 Substituting into formulae

1 The formula to work out the shaded area, A, of this shape is $A = \frac{1}{2}bh - lw$
Find the shaded area when $b = 12, h = 10, l = 5$ and $w = 4$.

E

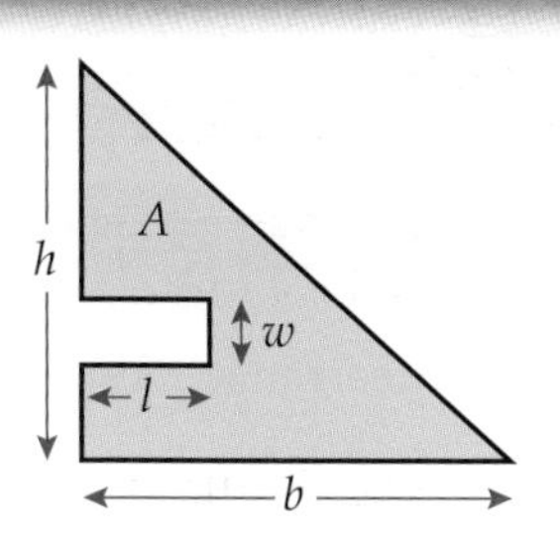

2 The formula for working out the perimeter of a regular octagon is $P = 8l$, where l is the length of each side. Work out the value of l when $P = 56$ cm.

3 An electricity company calculates electricity bills each quarter (3 months) using the formula:

E

$$C = 0.085U + 37.50$$

where C is the amount to pay in pounds, and U is the number of units of electricity used.
A customer uses 1450 units in April, 1000 units in May and 750 units in June.

a Work out the customer's bill for this quarter of the year.

b The customer would like his bill to be less than £250 per quarter.
How many units of electricity can he use, to have a bill of £250?

AO2

D

4 The value of a diamond, £V, can be worked out using the formula

$$V = 20M^2$$

where M is the mass of the diamond in grams.
Work out the value of a diamond that has a mass of 25 g.

5 The formula to work out surface area, SA, of a square-based pyramid is:

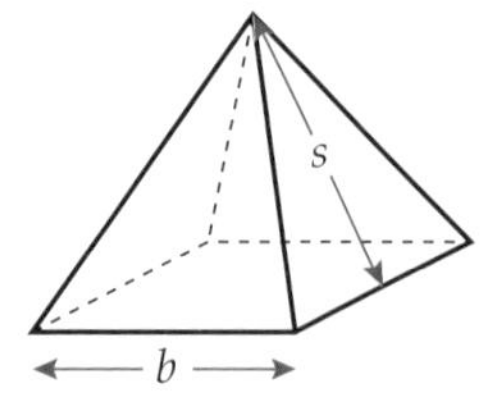

$$\text{SA} = b^2 + 2bs$$

where b is the base length and s is the perpendicular height of a triangular face.
Use the formula to work out the surface area of a square-based pyramid when $b = 5$ cm and $s = 7$ cm.

6 To calculate the height, h, of the Great Pyramid in Giza, Egypt, explorers used the formula

$$h = \sqrt{s^2 - \tfrac{1}{4}b^2}$$

where b is the base length and s is the perpendicular height of one of the triangular faces of the pyramid.
Use the formula to work out the height of the Great Pyramid.
The base height is 230 m and the perpendicular height of one of the triangular faces is 185 m.

7 The perimeter of an arrowhead can be worked out using the formula

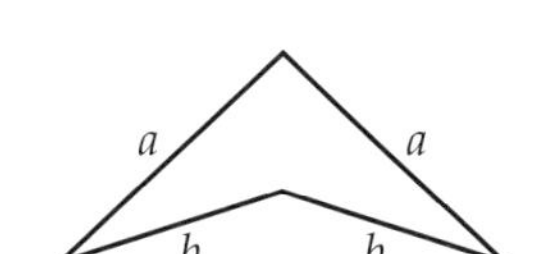

$$P = 2a + 2b$$

where a and b are the lengths of the sides.

Use the formula to work out

a a when $P = 32$ and $b = 6$

b b when $P = 40$ and $a = 13$.

8 Use the formula $x = 2\sqrt{\frac{E}{10}}$ to work out x when $E = 360$.

15.6 Changing the subject of a formula

C

1 Rearrange these formulae to make x the subject.

a $a = x + 6$ **b** $b = x - 6$ **c** $c = 5x$

d $d = tx$ **e** $e = 3x - 2$ **f** $f = 4x + 4$

g $g = 5x + 4t$ **h** $h = \frac{1}{2}x - 3$ **i** $i = 10(x + 4)$

B

2 Rearrange these formulae to make y the subject.

a $3(y - a) = 2 + a$ **b** $b = \frac{y}{3} - t$

c $8c = \frac{4y}{x} + t$ **d** $d = \frac{y^2}{t} - r$

e $e = \sqrt{y + x}$ **f** $f = 5\sqrt{y} - 3$

g $g = \sqrt{\frac{y}{t}}$ **h** $h = \frac{xy^2z}{3}$

3 Rearrange $5(x + y) = 2y - 8$ to make x the subject.

Links to:
Middle Student Book
Ch16, pp.259–274

Key Points

Credit **E**

When you buy on credit, you usually have to pay a deposit followed by a number of regular payments.

Calculations involving simple interest **E**

When you put money into a savings account in a bank or building society they pay you interest. If the interest is the same amount each year, it is called simple interest.

VAT **D**

Value Added Tax (VAT) is added to the price of items and services. Generally it is 17.5% in the UK.

Percentage increase and decrease **D**

Method A

1 Work out the value of the increase (or decrease).
2 Add to (or subtract from) the original amount.

Method B

1 Add the percentage increase to 100% (or subtract the percentage decrease from 100%).
2 Convert this percentage to a decimal.
3 Multiply it by the original amount.

Percentage profit or loss **C**

$$\text{Percentage profit (or loss)} = \frac{\text{actual profit (or loss)}}{\text{cost price}} \times 100\%$$

where the actual profit (or loss) is the difference between the cost price and the selling price.

Repeated percentage change **C**

When you invest money the interest is usually added to the amount of money invested in the first place. The interest that you get is calculated on the amount invested plus the interest already received. This is known as compound interest.

In section 8.1 you learned how to work out repeated percentage change using a calculator. To practise these without a calculator.

1 Work out the interest for the first year.
Add the interest to the amount you started with. This is the amount at the start of the next year.
2 Work out the interest on the new amount.
Add the interest to the amount at the start of that year.
3 Repeat for the number of years needed.

You can also use this method to work out a repeated percentage loss or depreciation. Work out the loss or depreciation each year and subtract it from the value at the start of the year.

Finding the original quantity **B**

To work out the original quantity when you are given the quantity after a percentage increase or decrease, use one of these methods.

Method A

1 Work out what percentage the figure you are given represents.
2 Divide by this percentage to find 1%.
3 Multiply by 100 to get the 100% figure.

Method B

1 Work out what percentage the figure you are given represents.
2 Divide by 100 to find the multiplier.
3 Divide the original amount by the multiplier.

16.1 Percentage increase and decrease

D

1 Increase the following quantities by 10%.

a £40 **b** 65 cm **c** 58 kg

2 Decrease the following quantities by 10%.

a £90 **b** 24 mm **c** 72 m*l*

3 **a** Increase £800 by 1%. **b** Decrease 120 km by 1%.
c Increase 48 m by 2%. **d** Decrease £64 by 8%.

4 Sandeep earns £1400 per month.
He is given a 5% pay rise.
How much does he now earn per month?

D

5 The cost of a holiday to Spain is £480.
Geraint books the holiday online and gets 15% off the price.
How much does Geraint pay for his holiday?

6 Harry wants 4 tonnes of top-soil for his garden.
He orders 10% more than he wants, just in case he doesn't have enough.
How much top-soil does he order?

1 tonne = 1000 kg

7 Alexi buys a dishwasher in a sale.
This is the price ticket on the dishwasher.
How much does Alexi pay for the dishwasher?

D

8 Roller-skates, priced at £60, are reduced by 20% in a sale.
After 2 weeks the roller-skates still haven't been sold,
so are reduced by a further 15%.
What is the final sale price of the roller-skates?

The second reduction is 15% of the first sale price, not the original price.

9 Ted weighed 80 kg before Christmas.
Over the Christmas period, he increased his weight by 15%.
After Christmas, he went on a diet and lost 15% of his new weight.
Does Ted now weigh the same as or more or less now than he did before Christmas?

A02

16.2 Calculations with money

1 Work out the VAT (17.5%) to be added to a CD player that costs £58.

2 Work out the VAT (17.5%) to be added to a DVD player that costs £180.

3 Work out the total cost of each of these.

a a slow-cooker that costs £42 + 17.5% VAT

b a bicycle that costs £480 + 17.5% VAT

c a gas bill that comes to £315 + 5% VAT

E

4 Reth buys some gardening equipment from a catalogue.
He pays £25 the first month and then £12 per month for the next 9 months.
How much does Reth pay in total?

5 Nadia buys a new kitchen that has a cash price of £5200.
The credit terms are a deposit of 15% and 24 monthly payments of £200.

a How much is the deposit?

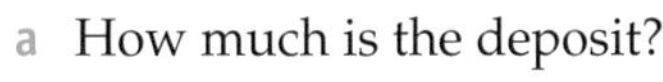

b What is the total of her monthly payments?

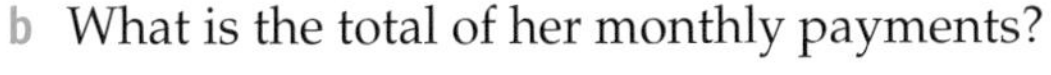

c What is the total credit price?

d How much more does it cost to buy on credit rather than pay by cash?

6 Ricardo puts £800 into a savings account. The interest rate is 3.5% per annum.
Work out the simple interest after 5 years.

7 The cash price for new kite-surfing equipment is £1200.
The credit terms for the kite-surfing equipment are 20% deposit plus 18 monthly payments of £60.
What is the difference between the cash price and the credit price?

E

8 Sean and Dave both have £1200 to invest.
They both invest their money into savings accounts that pay simple interest.
Sean's account gives him 5% per annum.
Dave's account gives him 7% for the first year, then 3% per annum for the following years.
They both invest their money for five years.
Who will have more money at the end of the five years?

AO2

16.3 Percentage profit and loss

C

1 Sally buys a painting for £65 and sells it for £91.
What is her percentage profit?

2 Carolyn buys a house for £112 000. She sells it five years later for £196 000.
What is her percentage profit?

3 Jaime bought a running machine for £950. He later sold it for £570.
What was his percentage loss?

4 Jim sold his laptop for £135. He had paid £675 for it.
What was his percentage loss?

C

5 Donna buys cushions for £12 and sells them for £30.
Sheila buys cushions for £20 and sells them for £48.
Who makes the larger percentage profit?

6 Henry bought a saxophone for £235. He sold it two years later for £188.
Torin bought a saxophone for £188. He sold it three years later for £141.
Who has the larger percentage loss?

C

AO3

7 Podraig bought a rare LP record for £75 and sold it again for £100.
Podraig says, '£25 out of the £100 is profit so I've made 25% profit on the LP.'
Is Podraig correct? Show working to explain your answer.

16.4 Repeated percentage change

C

1 Bob invests £120 at rate of 4% per annum compound interest.
How much will he have at the end of three years?

2 Lauren invests £800 at a rate of 5.5% per annum compound interest.
How much will she have at the end of three years?

3 The number of hedgehogs is going down by 8% each year.
In one area, there are estimated to be 800 hedgehogs.
Estimate the number of hedgehogs in this area in two years' time.

4 The value of a static caravan depreciates by 20% each year.
John buys a static caravan for £36 000.
How much will it be worth after three years?

5 The Havens football team has 220 supporters. The number of supporters is increasing at a rate of approximately 5% per annum.
The Docks football team has 185 supporters. The number of supporters is increasing at a rate of approximately 12% per annum.
After three years, which team has more supporters?

> Remember to round your answer to the nearest whole number. You can't have a fraction of a supporter!

6 Grassholm Island in Pembrokeshire has one of the largest gannet colonies in the world. In 2009 there were approximately 37 000 breeding pairs.
The number of gannets is increasing at an average of 2% per annum.
In which year will the number of breeding pairs be expected to reach 40 000?

7 Bacteria increase on average by 100% every 20 minutes on an uncooked chicken at room temperature.
There are 5000 bacteria at 10 am. How many bacteria will there be at midday?

16.5 Reverse percentages

Sale!
20% off
Washing machine
Now only £440

1 A washing machine is on sale.
This is the price ticket on the washing machine.
How much did the washing machine cost before the sale?

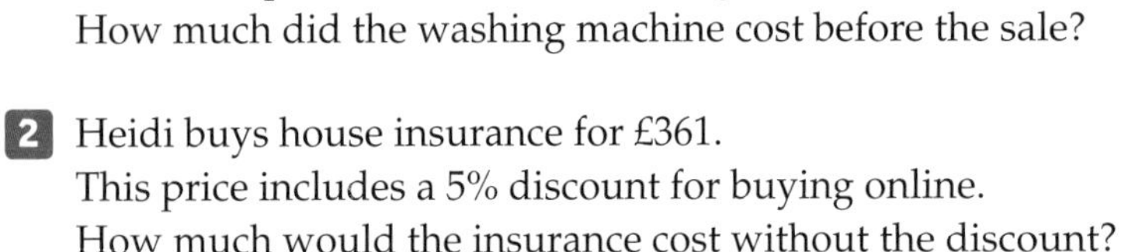

2 Heidi buys house insurance for £361.
This price includes a 5% discount for buying online.
How much would the insurance cost without the discount?

3 Bruce has an antique clock. It is valued at £900.
This is 25% more than he paid for it.
How much did Bruce pay for the clock?

4 Ross decides to go on a diet.
He now weighs 82.8 kg, which is 15% more than his ideal weight.
What is his ideal weight?

5 Sandra buys a coat for £35 in a sale.
The coat had a discount of 20%.
Sandra says, 'The original price of the coat was £42.'
Explain why Sandra is wrong.

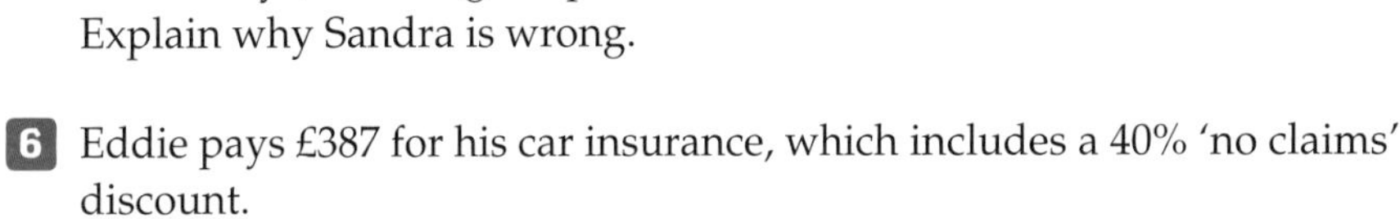

6 Eddie pays £387 for his car insurance, which includes a 40% 'no claims' discount.
Jim pays £330 for his car insurance, which includes a 50% 'no claims' discount.
Eddie says, 'My insurance was cheaper than Jim's before the discount.'
Is Eddie correct? Show working to support your answer.

17 Sequences and proof

Links to:
Middle Student Book
Ch17, pp.278–296

Key Points

Sequences E

A sequence is a set of numbers in a given order.

To find the next term in the sequence, look at the differences between consecutive terms. Use these to find the term-to-term rule.

The nth (general) term D

The nth term of a sequence can be used to find any term in the sequence.

Finding the nth term of a linear sequence C

To find the nth term of a linear sequence, first look at the difference between consecutive terms. This tells you the multiplier for n in the nth term.

Then compare the terms in the sequence with the multiples of the difference. This tells you what has to be added or taken away to match the nth term to the sequence.

The nth term of a simple quadratic sequence C

In a quadratic sequence, the difference between consecutive terms is *not* constant.

But the second difference *is* constant.

The nth term of a quadratic sequence includes an n^2.

Proof E D C

A proof uses logical reasoning to show something is true.

Using counter-examples C

A counter-example is an example which shows that a statement is false.

You can disprove a statement if you can find one example that doesn't fit it.

17.1 Number patterns

1 For each sequence, first find the next three terms, and then write down the term-to-term rule.

a $-10, -7, -4, -1, \ldots$

b $15, 7, -1, -9, \ldots$

2 The first term of a sequence is 17. The term-to-term rule is subtract 5.
How many positive terms are there in the sequence?

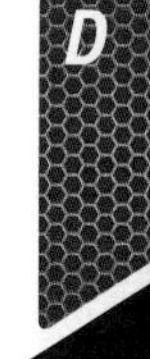

3 **a** Write down the first five terms of two different sequences that start 1, 3, …

b Describe in words the rules for continuing both sequences.

c Here is a sequence with some terms missing.

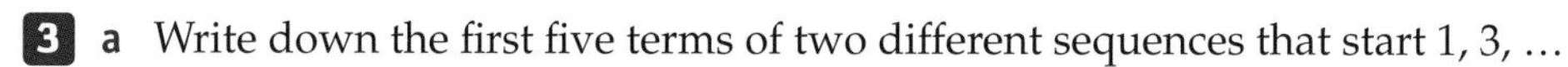

1, 3, □, □, 15, 21, …

What is the 8th term in this sequence?

4 For each sequence, work out the next two terms and describe the pattern of differences.

a 10, 12, 16, 22, …

b 50, 45, 35, 20, …

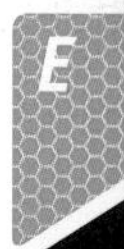

5 The first four terms of a sequence are 80, 70, 59, 47.
What is the value of the first term in the sequence which is less than zero?

6 The first four terms of a sequence are 3, 15, 29, 45.
What is the value of the first term in the sequence which is greater than 100?

E

7 Lin has £500 in a savings account. To increase her savings she decides to pay an extra amount per month into this account. The table shows the amount in her account at the end of every month.

	May	June	July	Aug
Amount in account (£)	500	525	575	650

The amount Lin pays into her account continues to increase in the same way. At the end of which month will Lin have more than £2000 in her account?

AO3

17.2 Rules for sequences

E

1 The general term of a sequence is $2n + 3$.

a Find the first five terms.

b Find the 20th term.

c Find the 100th term.

d What is the difference between consecutive terms?

E

2 The general term of a sequence is $n - 3$.
How many negative terms are there in the sequence?

3 The general term of a sequence is $3n + 4$.
Which term is the first term larger than 40?

AO2

C

4 Find the nth term of each sequence.

a 4, 8, 12, 16, 20, … b 4, 6, 8, 10, 12, … c 10, 13, 16, 19, 22, … d 25, 20, 15, 10, …

5 The first four terms of a sequence are 4, 11, 18, 25.

a Find the nth term of the sequence.

b Use the nth term to find the i 20th term ii 200th term

17.3 Using the nth term

C

1 Find the 20th term of this sequence.

2, 5, 8, 11, …

2 Find the 25th term of this sequence.

10, 16, 22, 28, …

3 Find the 30th term of this sequence.

30, 22, 14, 6, …

4 A sequence begins 10, 14, 18, 22, …
One of the terms in the sequence is 118.
Which number term is this?

5 Here is a sequence.

3, 11, 19, 27, …

a One of the terms in the sequence is 251.
Which number term is this?

AO2

b Explain why 342 cannot be a term in this sequence.

6 The first four terms of a sequence are 7, 11, 15, 19.
Find the term that is closest to 100.

7 The first four terms of a sequence are 8, 5, 2, -1.
Find the term that is closest to -50.

C

AO3

17.4 Sequences of patterns

E

1 Here is a sequence of dot patterns.

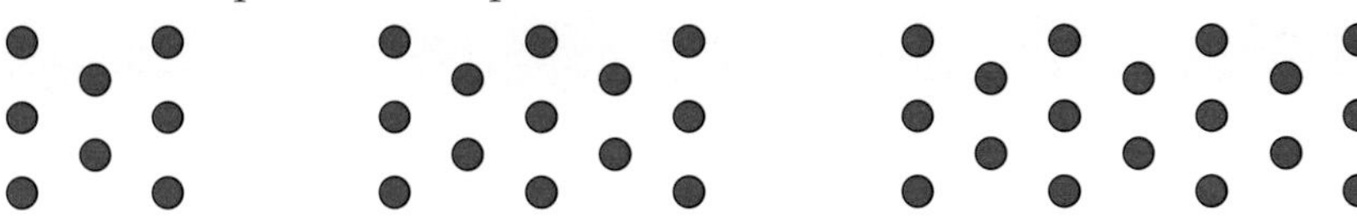

a Without drawing the pattern, work out how many dots are in pattern 4.

b Explain how you worked out your answer to part a.

2 Here is a sequence of dot patterns.

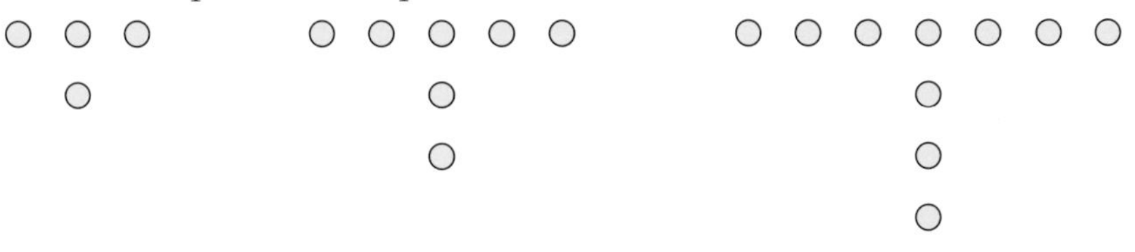

a Without drawing the pattern, work out how many dots are in pattern 4.

b Explain how you worked out your answer to part a.

AO2

C

3 A farmer uses post and rail fencing, as shown in the diagram.

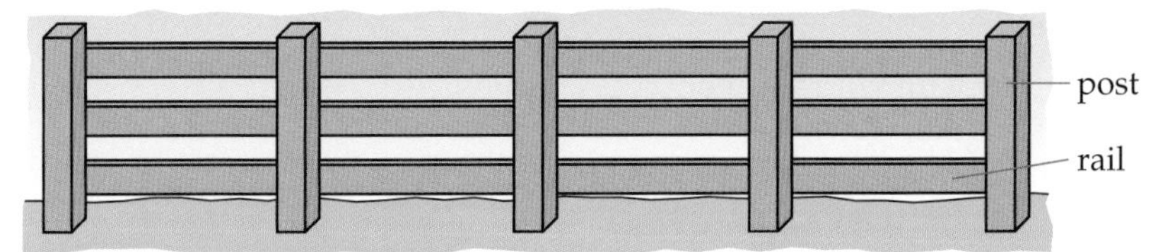

a Copy and complete the table.

Number of sections	1	2	3	4	5
Number of posts	2	3			
Number of rails	3				

b How many posts does the farmer need for n sections of fencing?

c How many rails does the farmer need for n sections of fencing?

d How many posts and rails does the farmer need for 35 sections of fencing?

4 In a restaurant, tables can be pushed together for larger parties, as shown in the diagram.

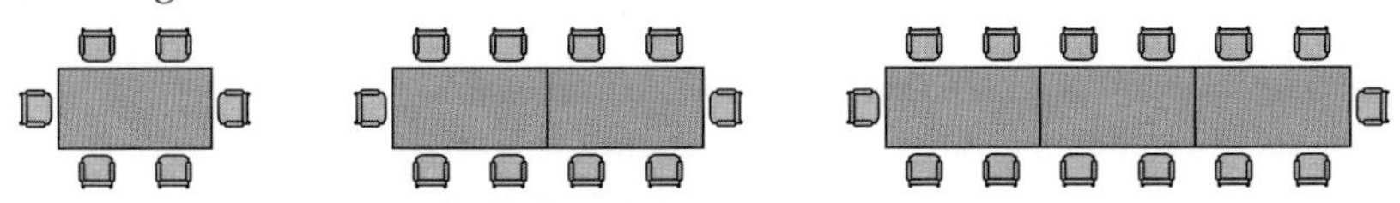

a Copy and complete the table.

Number of tables	1	2	3	4	5
Number of people	6				

b How many people can sit at n tables?

AO2

C

5 The farmer in Q3 has a huge pile of posts but only has 40 rails.
The farmer makes a fence as long as possible.
How many rails will be left over?

6 The manager at the restaurant in Q4 has 52 chairs altogether.
He makes a row of as many tables as he can.
How many chairs will he have left over?

AO2

17.5 Quadratic sequences

D

1 The general term of a sequence is $n^2 + 5$.

a Work out the first five terms of the sequence.

b What is the 20th term of the sequence?

2 The nth term of a sequence is $2n^2$.

a Work out the first five terms of the sequence.

b What is the 12th term of the sequence?

D

3 Shamine says, 'In the sequence with nth term $= n^2 - 2$, all the terms are positive.'
Is Shamine correct? Explain your answer.

4 The general term of a sequence is $\frac{n^2}{2}$.
Which term has value 32?

5 A sequence has general term $3n^2 + 1$.
Is 121 a number in the sequence? Explain your answer.

6 Copy and complete this quadratic sequence.

1, 7, ☐, ☐, 49, 71, 97

AO2

17.6 The nth term of a simple quadratic sequence

C

1 Find the nth term and the 10th term of each sequence.

a 1, 4, 9, 16, …

b 5, 8, 13, 20, …

c 12, 15, 20, 27, …

d 33, 36, 41, 48, …

> In parts b, c and d, compare each sequence with the sequence in part a.

2 Find the nth term and the 10th term of each sequence.

a 2, 8, 18, 32, …

> In part a, compare each sequence with the sequence in Q1a.

b 4, 10, 20, 34, …

> In parts b and c, compare each sequence with the sequence in part a.

c 1, 7, 17, 31, …

d −4, −1, 4, 11, …

17.7 Proof

1 I think of a positive integer, multiply it by 2 and add 3.
  a What type of number (odd or even) will I get?
  b Explain how you know.

2 I think of a positive integer greater than 5, take away 1 and multiply the result by 2.
  a What type of number (odd or even) will I get?
  b Explain how you know.

3 Lily is exactly two years older than her sister Holly.
Lily says, 'If I add my age to Holly's age, the total will always be an even number.'
Explain why Lily is correct.

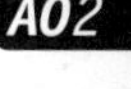

4 n represents an integer.
  a Is $n + (n + 3)$ an odd number or an even number?
  b Explain how you know.

5 n is an odd number larger than 1.
Explain why $n^2 - 1$ is always an even number.

6 Show that the sum of three consecutive odd numbers is always an odd number.

7 Show that the sum of four consecutive odd numbers is always an even number.

17.8 Using counter-examples

1 Give a counter-example to show that each of these statements is false.
  a When you cube a number, the answer is always larger than the number you started with.
  b When you multiply an integer by a fraction, the answer is always smaller than the integer you started with.

2 Alan says, 'I am thinking of a number. When I round my number to the nearest 100, I get an answer of 400.' Moira says that Alan's number must be bigger than 390.
Give a counter-example to show that Moira is wrong.

3 China says, 'If x and y are odd numbers, then $x^2 + y^2$ is also an odd number.'
Give a counter-example to show that China is wrong.

4 Rio says, 'The sum of two consecutive square numbers is always even.'
Explain why Rio is wrong.

5 Ashley says, 'If x is an integer, then $x^2 + 3$ is always odd.'
Explain why Ashley is wrong.

6 Brian says, 'If x is an integer, then $x^3 - 1$ is always even.'
Explain why Brian is wrong.

18 Linear graphs

Links to: Middle Student Book Ch18, pp.297–321

Key Points

Mid-point of a line segment (D C)

Mid-point $(x, y) = \left(\frac{x_1 + x_2}{2}, \frac{y_1 + y_2}{2}\right)$

where (x_1, y_1) and (x_2, y_2) are the coordinates of the end-points.

Straight-line graphs (E D C B)

A straight-line graph parallel to the x-axis has equation y = a number.

A straight-line graph parallel to the y-axis has equation x = a number.

Straight-line graphs have equations of the form $y = mx + c$.

The gradient (slope), m, of a straight line measures how steep it is.

Gradient, $m = \frac{\text{change in } y}{\text{change in } x}$

The value of c is the y-intercept.

Conversion graphs (E)

A conversion graph converts values from one unit to another.

Distance-time graphs (E D C)

A distance–time graph represents a journey. The x-axis (horizontal) represents the time taken. The y-axis (vertical) represents the distance from the starting point.

The average speed of a journey can be worked out using

Average speed $= \frac{\text{total distance}}{\text{total time}}$

A straight-line graph shows that the rate of change is steady.

A curved graph shows that the rate of change varies.

The steeper the line, the faster the rate of change.

Simultaneous equations (B)

You can solve a pair of simultaneous equations by drawing their graphs on the same set of axes.

The point on the graph where the two lines intersect (cross) is their solution.

Inequalities (B)

Inequalities can be shown as a shaded region on a graph.

For the inequalities $\leqslant$ and $\geqslant$, the boundary is a solid line.

For the inequalities $<$ and $>$, the boundary is a dashed line.

18.1 Mid-point of a line segment

D

1 For each line segment

a write down the coordinates of the end-points

b work out the coordinates of the mid-point.

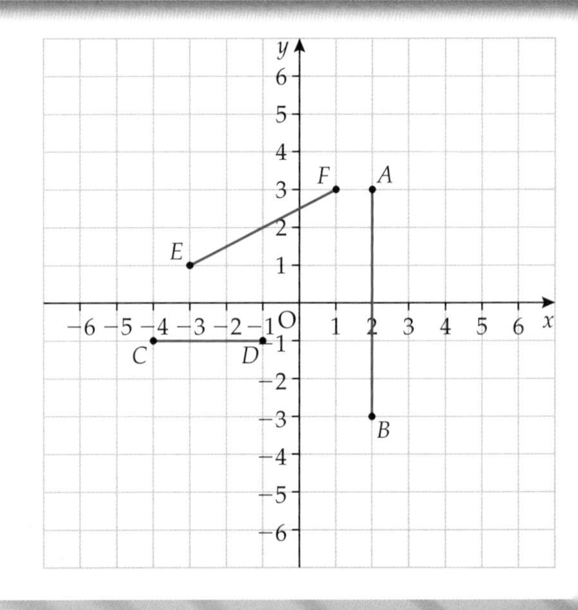

2 Work out the mid-points of these line segments.

a GH: $G(-1, -4)$ and $H(3, -4)$

b IJ: $I(7, 2)$ and $J(3, -2)$

c KL: $K(-3, 1)$ and $L(0, 4)$

C

3 $MNPQRS$ is a shape with coordinates $M(-3, 2)$, $N(3, 4)$, $P(4, -2)$, $Q(0, -2)$, $R(0, -4)$ and $S(-4, 0)$.

a Work out the coordinates of the mid-point of the side MN.

b Work out the coordinates of the mid-point of the side PQ.

c Work out the coordinates of the mid-point of the side RS.

d What shape do you get if you join the mid-points of sides MN, PQ and RS?

C AO2

4 The point T has coordinates $(-3, 4)$.
The mid-point of the line segment TU has coordinates $(0, 2\frac{1}{2})$.
Work out the coordinates of the point U.

C AO3

18.2 Plotting straight-line graphs

1 Write down the equations of each line on the grid that is

a parallel to the x-axis

b parallel to the y-axis.

E

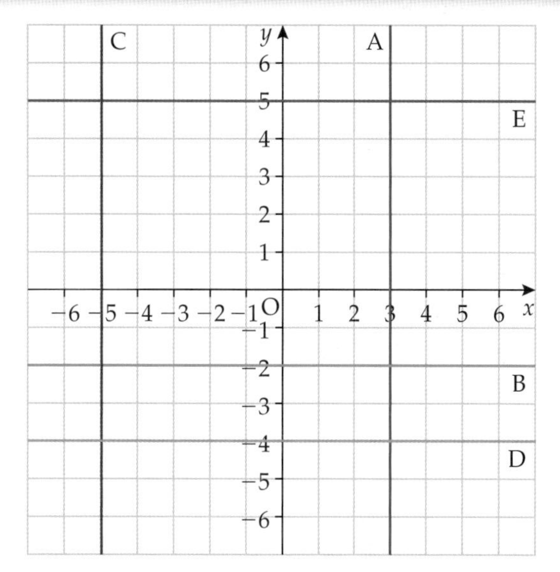

2 Draw and label these graphs.

a $y = -1$ b $x = 4$ c $x = 0$ d $y = 0$

3 Draw the graph of $y = 3x - 2$ for $-3 \leqslant x \leqslant 3$.

D

Start by drawing a table of values from $x = -3$ to $x = 3$ and work out the y-coordinate that goes with each x-coordinate.

4 a Copy and complete this table of values for $y = 2 - 2x$.

x	−2	−1	0	1	2
y	6				−2

b Draw the graph of $y = 2 - 2x$.

c Draw the line $x = 1$ on your graph.

d P is the point where the two lines cross.
Mark the point P and write down its coordinates.

D AO2

D AO3

5 'The lines $y = 4x - 3$ and $x = -1$ cross at the point $(-1, -8)$.'
Is this statement correct? Show working to support your answer.

C

6 a Copy and complete this table of values for $x + 2y = 8$.
b Draw the graph of $x + 2y = 8$.

x	0	
y		0

C AO2

7 Draw the graph of $2y + 3x = 18$.

Substitute $x = 0$ then $y = 0$ into the equation.

18.3 Equations of straight-line graphs

D AO2

1 a On the same coordinate grid, draw the graphs of
i $y = \frac{1}{2}x$ ii $y = 2x + 1$ iii $y = \frac{1}{2}x - 2$
b Which line is the steepest?
c How can you tell which line is steepest from the equations?
d Which lines are parallel to each other?
e How can you tell which lines are parallel from the equations?

C

2 Work out the gradients of lines A and B.

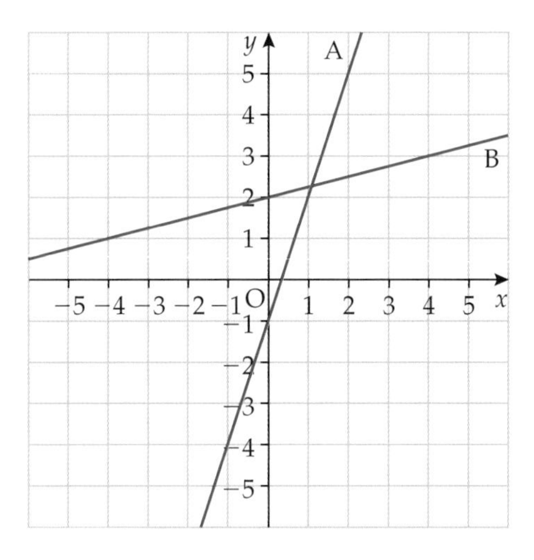

3 Write down the gradient of these straight lines.
a $y = 5x - 1$ b $y = 4 - \frac{1}{2}x$ c $y = -3x + 3$

4 Which of these straight lines are parallel to $y = -2x + 4$?
A $y = -2x - 3$ **B** $y = 1 - 2x$ **C** $y = 2x - 8$

5 Write down the y-intercepts of these lines:
a $y = \frac{1}{2}x + 4$ b $y = 3\frac{1}{2} + x$ c $y = \frac{1}{4}x - 5$

6 Write the following equations in the form $y = mx + c$.
a $3y = 9x + 12$ b $y - 4 = 4x$ c $5y + 15x = 45$

C AO2

7 Without plotting these straight lines, identify the ones parallel to the line $y = 2x - 3\frac{1}{2}$.
A $2y = 4x - 9$ **B** $3y - 6x = 3$ **C** $2y + 4x = 7$
D $5y = 15x + 10$ **E** $2x = 8 + y$ **F** $2x = 4y + 4$

8 A line has gradient 4 and passes through the point (0, 7).
Work out the equation of the line.

B

9 A line passes through the points (−2, 1) and (4, 4). Work out

a the gradient of the line   b the equation of the line.

B

10 A straight line parallel to $y = \frac{1}{2}x - 2$ passes through the point (0, 3).
What is the equation of this line?

When two lines are parallel, their gradients are equal.

18.4 Conversion graphs

E

1 This is a conversion graph between miles and kilometres.

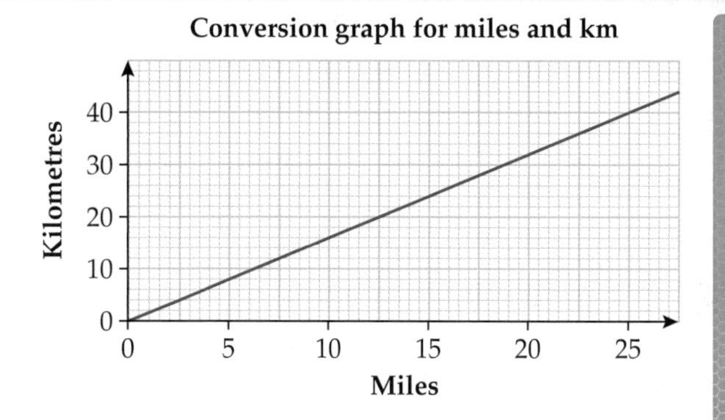

a Use the graph to convert
  i 10 miles into km
  ii 32 km into miles.

b Use your answers to part a to convert
  i 70 miles into km
  ii 128 km into miles.

2 The conversion rate from pounds sterling (£) to Barbados dollars (Bds$) is £1 = 3.20 Bds$.

a Copy and complete this conversion table between pounds and Bds$.

£ (x)	0	5	10
Bds$ (y)			32

b Draw a conversion graph with x-values from £0 to £10 and y-values from 0 Bds$ to 32 Bds$.

c Use your graph to convert
  i £3 to Bds$
  ii 20 Bds$ to pounds.

d Use your answers to part c to find the value of
  i £33 in Bds$
  ii 80 Bds$ in pounds.

AO2

18.5 Real-life graphs

E

1 Anthony is going by coach on holiday.
The coach travels a distance of 80 km in the first hour and 90 km in the second.
The coach then stops for a break for $\frac{3}{4}$ of an hour.
The coach travels the final 100 km in $1\frac{1}{2}$ hours.

a Draw a distance–time graph to show Anthony's journey.

b During what section of the journey was the coach travelling at its fastest?
Give a reason for your answer.

D

2 Every Saturday Abu walks into town.
He stops on the way home to buy fish and chips.
The graph shows his journey one Saturday.

Abu's journey to town

a How far has Abu gone in the first $\frac{1}{2}$ hour?

b How far is it from Abu's house to town?

c How long did Abu stay in town?

d How long did it take Abu to buy fish and chips? How do you know?

e Abu left his house at 1 pm. What time did he get home?

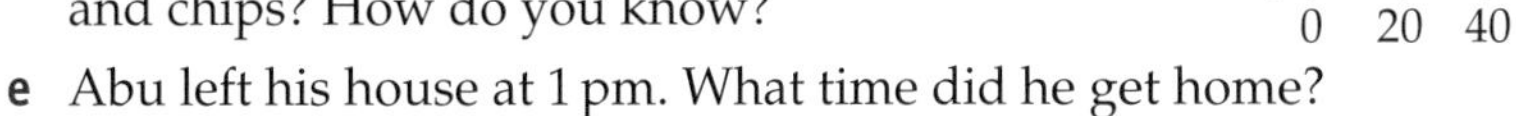

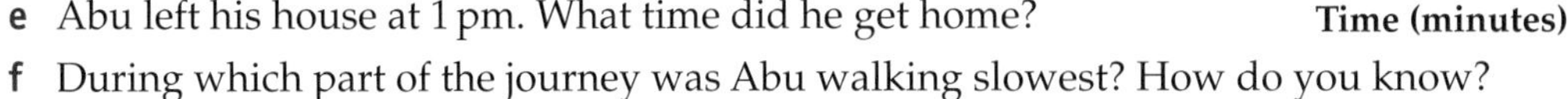

f During which part of the journey was Abu walking slowest? How do you know?

g Work out his average speed for the journey home.

3 Sam is driving from Edinburgh to Aberdeen for an interview.
She sets off at 8 am and travels 90 km in the first hour.
She travels the next 60 km in half an hour. She then stops for a 20-minute break.
The remaining 50 km takes her 20 minutes.

a Draw a distance–time graph for Sam's journey.

b What time does she arrive in Aberdeen?

c During which part of the journey was Sam travelling the fastest?

d What was her speed during the fastest part of the journey?

e What was her average speed for the whole journey?

C

4 Work out the speed of the journey shown by this distance–time graph.

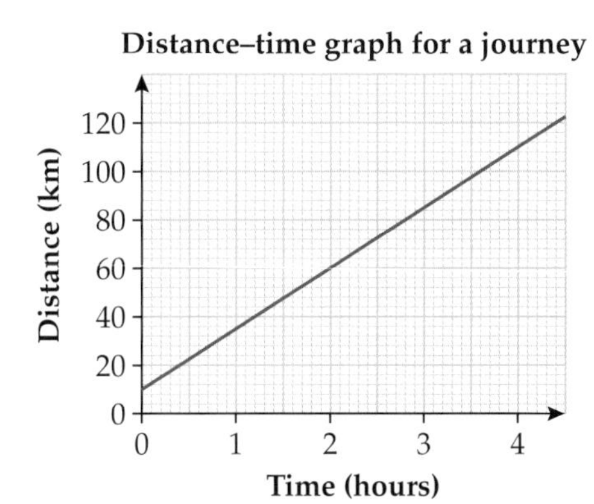

C

5 Water is poured at a steady rate into these containers.

A

B

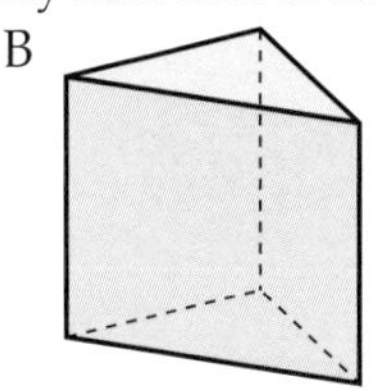

C

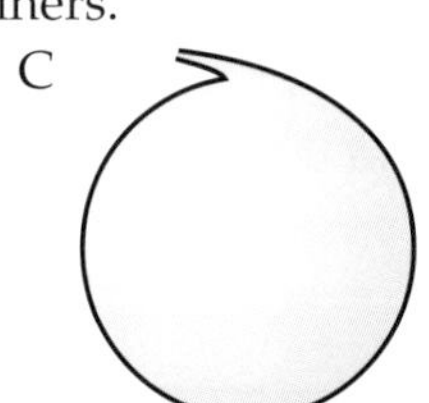

D

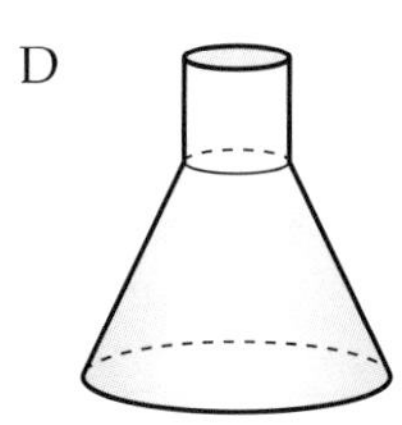

The depth of water in the containers is measured over time and a graph plotted.

i

ii

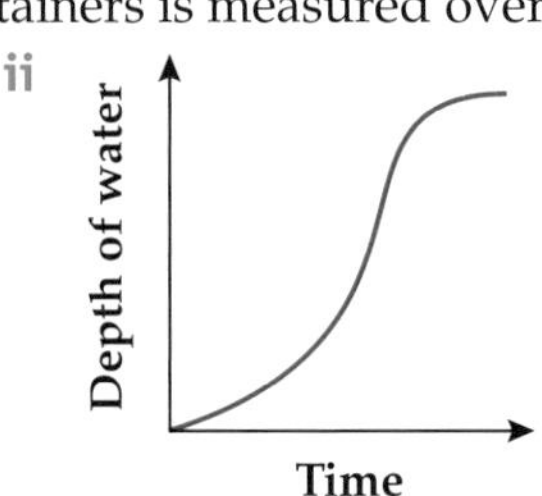

iii

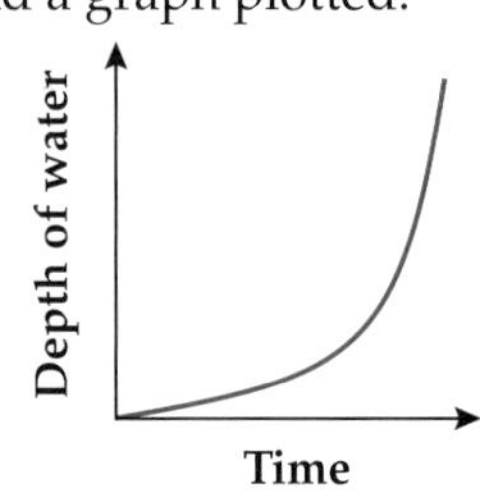

a Match the containers to the graphs.

b One container has not been matched. Which one is it?

c Sketch a graph for this container.

AO3

18.6 Using graphs to solve simultaneous equations

B

1 Solve each pair of simultaneous equations graphically.

a $x + y = 10$
$2x + y = 14$

b $4x + y = 12$
$3x - y = 2$

B

2 Josh starts work at a restaurant.
This is the 'specials' menu board.

Specials
Lasagne £
Curry & chips £9
Fish & chips £11
Beef pie £
Vegetable risotto £8

Two of the prices have accidentally been rubbed out.
Josh knows that one group of people ordered 3 lasagnes and 2 beef pies and it cost £48.
Another group of people ordered 2 lasagnes and 1 beef pie and it cost £28.
Josh writes down these two equations.

$3x + 2y = 48$
$2x + y = 28$

a Use a graphical method to solve the simultaneous equations.

b What is the price of a lasagne?

c What is the price of a beef pie?

AO2

18.7 Using graphs to solve inequalities

B

1 Use inequalities to describe the shaded regions on these graphs.

a
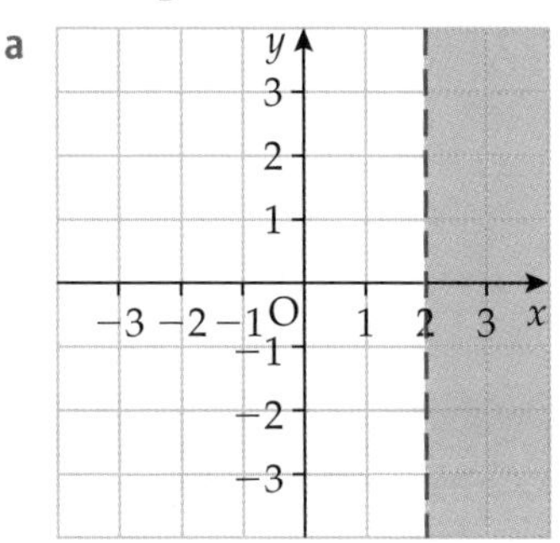

b
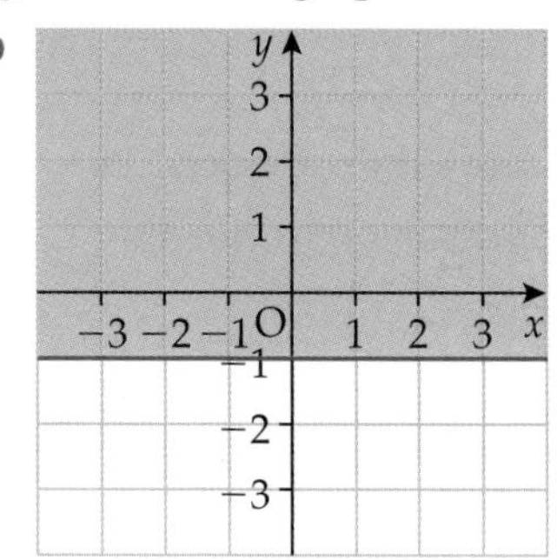

2 Draw graphs to show the region that satisfies each inequality.

a $x \leqslant 3$ b $y > 1$ c $y \geqslant x$ d $y < x + 1$

3 Sketch the region defined by these three inequalities.

$x \geqslant -1$ $y \geqslant -2$ $x + y < 4$

Mark the region with an 'R'.

19 Quadratic equations

Links to:
Middle Student Book
Ch19, pp.322–332

Key Points

Difference of two squares B

The difference of two squares is any expression which can be written in the form

$$(x - b)(x + b)$$

It will always be of the form $x^2 - b^2$.

Factorising quadratic expressions B

To factorise an expression of the form $x^2 + bx + c$ you must find two numbers whose product is c and whose sum is b.

For example, $x^2 + 5x + 6 = (x + 2)(x + 3)$

Solving quadratic equations by rearranging B

You can solve some quadratic equations by rearranging them to make the unknown the subject.

For example

$$6y^2 - 24 = 0$$
$$6y^2 = 24$$
$$y^2 = 4$$
$$y = \pm 2$$

Solving quadratic equations by factorising B

You can solve some quadratic equations by rearranging them so that all the terms are on one side, and then factorising. For example

$$x^2 - 5x + 6 = 0$$
$$(x - 2)(x - 3) = 0$$
$$x - 2 = 0 \quad \text{or} \quad x - 3 = 0$$

So $x = 2$ or $x = 3$

19.1 Factorising the difference of two squares

B

1 a Copy and complete this statement.

$$x^2 - 144 = (x + \square)(x - \square)$$

b Check your answer to part **a** by expanding the brackets.

2 Factorise the following expressions.

a $a^2 - 81$ b $b^2 - 400$ c $c^2 - 36$

3 Factorise the expression $x^2 - 10\,000$.

B A02

4 Andy thinks of a number, squares it and subtracts 100.

a Write down an algebraic expression to decribe this.

b Factorise your answer to part **a**.

19.2 Factorising quadratics of the form $x^2 + bx + c$

B

1 Copy and complete the following factorisations.

a $x^2 + 10x + 24 = (x + 6)(x + \square)$
b $x^2 + 11x + 18 = (x + \square)(x + 2)$
c $x^2 - 18x + 32 = (x - 16)(x - \square)$
d $x^2 - 13x + 36 = (x - \square)(x - 9)$
e $x^2 - 20x - 44 = (x - 22)(x + \square)$
f $x^2 - 4x - 32 = (x - \square)(x + 4)$
g $x^2 - 12x - 28 = (x + 2)(x - \square)$
h $x^2 + 5x - 24 = (x + \square)(x - 3)$

2 Factorise the following expressions.

a $x^2 + 8x + 12$ b $x^2 + 9x + 18$ c $x^2 - 15x + 36$
d $x^2 - 12x + 32$ e $x^2 - 7x - 44$ f $x^2 - 2x - 63$
g $x^2 + 3x - 28$ h $x^2 + 2x - 24$ i $x^2 - 6x - 27$

3 Factorise the following expressions.

a $a^2 + 2a + 1$ b $b^2 + 12b + 36$ c $c^2 - 18c + 81$ d $d^2 - 20d + 100$

B

19.3 Solving quadratic equations

1 Solve the following equations by rearranging.

a $a^2 = 25$ b $b^2 + 9 = 45$ c $c^2 - 12 = 132$

d $5d^2 = 20$ e $3e^2 + 7 = 115$ f $10f^2 - 8 = 82$

g $\frac{g^2}{3} = 27$ h $5(x + 1)^2 = 80$ i $18 - 3x^2 = 6$

B

2 An estimate for the surface area of a sphere is given by the formula

$A = 12r^2$

where r is the radius of the sphere.
A sphere has a surface area of $432\,cm^2$.
Work out an estimate for the value of r.

B

3 A rectangular piece of paper is twice as long as it is wide.
The area of the piece of paper is $242\,cm^2$.

a By using width $= x$, form an equation involving the area of the paper.

b Solve the equation in part **a** to find x.

AO2

4 A plank of wood is 20 times longer than it is wide.
The area of the top face of the plank is $2880\,cm^2$.
What is the width of the plank of wood?

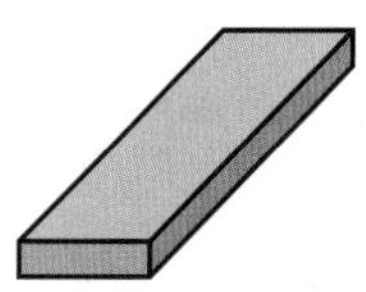

B

AO3

5 Solve the following quadratic equations by factorising.

a $x^2 + 12x = 0$ b $x^2 - 6x = 0$ c $x = 3x^2$

d $x^2 + 7x + 12 = 0$ e $x^2 + 4x - 12 = 0$ f $x^2 - x - 90 = 0$

g $x^2 + 14x = -24$ h $x^2 - 12 = -11x$ i $72 + x = x^2$

B

19.4 Writing and solving quadratic equations

1 Asha is 5 years older than her brother. The product of their ages is 84.
Use x to represent Asha's age.

a Write down an algebraic expression for her brother's age.

b Write down and simplify an algebraic expression for the product of their ages.

c Form and solve an algebraic equation to find the value of x.

d How old is Asha's brother?

B

2 Last year a farmer put a fence around one of her fields.
She remembers that the length of the field is three times the width. She also knows that the field has an area of $67\,500\,m^2$. Unfortunately she can't remember the dimensions of the field.
Form and solve a quadratic equation to work out the dimensions of the field.

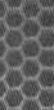

AO2

3 I think of a negative number, square it and add it to twice the original number.
My result is 35.
What number did I think of?

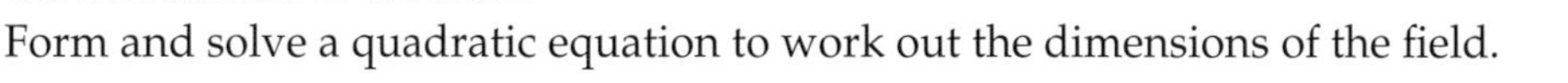

4 The product of two consecutive even numbers is 168.
Form and solve a quadratic equation to find the two numbers.

If the first even number is x, what is the next even number?

B

AO3

Links to:
Middle Student Book Ch20, pp.335–339

This section revises the number skills that you will need to use in Unit 3.

1 All except one of these fractions can be put into pairs of equivalent fractions. Which one of these fractions has no equivalent partner?

$\frac{8}{16}$ $\frac{2}{3}$ $\frac{1}{2}$ $\frac{5}{8}$ $\frac{10}{12}$ $\frac{3}{7}$ $\frac{14}{15}$ $\frac{15}{35}$ $\frac{10}{15}$ $\frac{5}{6}$ $\frac{28}{30}$

2 Use the numbers from the cloud to complete these statements

6 10 4 2 7 3 8 5

a The reciprocal of 8 is $\frac{1}{?}$.
b The reciprocal of $\frac{1}{2}$ is ?
c The reciprocal of $\frac{3}{4}$ is $\frac{?}{?}$.
d The reciprocal of $1\frac{1}{5}$ is $\frac{?}{?}$.
e The reciprocal of 0.1 is ?
f The reciprocal of 3.5 is $\frac{?}{?}$.
g Which number from the cloud have you used twice?

3 At a rugby match there were 1200 supporters altogether.
The table shows how many of the supporters were men, women, boys and girls.

	Men	Women	Boys	Girls
Number of supporters	480	320	240	160

a What percentage of the supporters were
i men ii women?
b Write the ratio of girls : women in its simplest form.
c Write the ratio of women : men in its simplest form.

4 Adam has these two number cards.

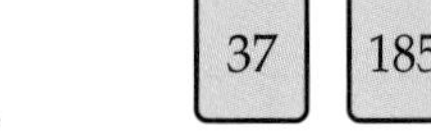

a Work out the sum of the two numbers.
b Work out the difference of the two numbers.
c Work out the product of the two numbers.

5 Work out the missing number in each of these calculations.

a $4 \times 5 + 11 = \square$
b $15 - 6 \times 2 = \square$
c $(24 + 6) \div 5 = \square$
d $8 + 4 \times \square = 16$
e $12 \times 2 - \square = 19$
f $(\square + 8) \times 5 = 60$
g $5 + 6^2 = \square$
h $8^2 - \square = 58$
i $4^2 \times \square = 64$

6 Copy and complete the table.
Write the fractions in their simplest form.

Percentage				60%	
Decimal		0.05			0.12
Fraction	$\frac{7}{10}$		$\frac{2}{5}$		

7 Work out

a 48×0.2
b 1.8×3.9
c $26.6 \div 7$
d $6.96 \div 0.12$

8 Choose the correct answer A, B or C for each of these.

	A	B	C
a 13.89 rounded to 1 d.p. is	**A** 14	**B** 13.8	**C** 13.9
b 0.067 2 rounded to 2 d.p. is	**A** 0.067	**B** 0.07	**C** 0.67
c 5.389 9 rounded to 3 d.p. is	**A** 5.39	**B** 5.400	**C** 5.390
d 1299 rounded to 1 s.f. is	**A** 1000	**B** 1	**C** 1300
e 985 rounded to 2 s.f. is	**A** 1000	**B** 990	**C** 980
f 0.502 16 rounded to 3 s.f. is	**A** 0.502 2	**B** 0.502	**C** 0.50

Links to:
Middle Student Book
Ch21, pp.340–356

Key Points

Angle properties

Angles on a straight line add up to 180°.

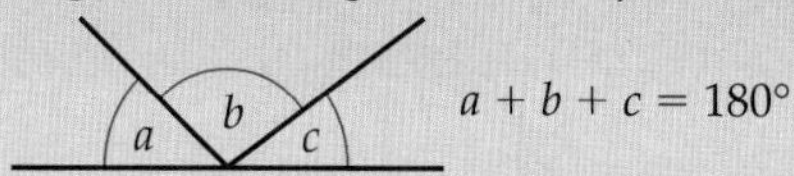

Angles around a point add up to 360°.

$d + e + f + g + h = 360°$

Vertically opposite angles are equal.

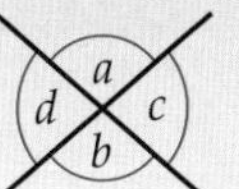

$a = b, c = d$

Corresponding angles are equal.

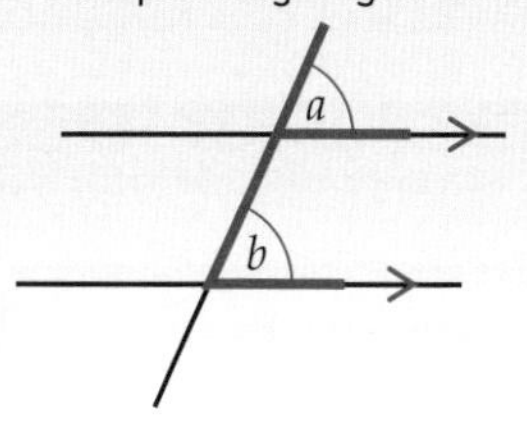

$a = b$

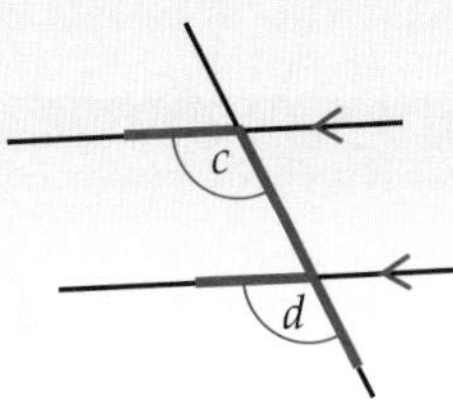

$c = d$

E D Alternate angles are equal.

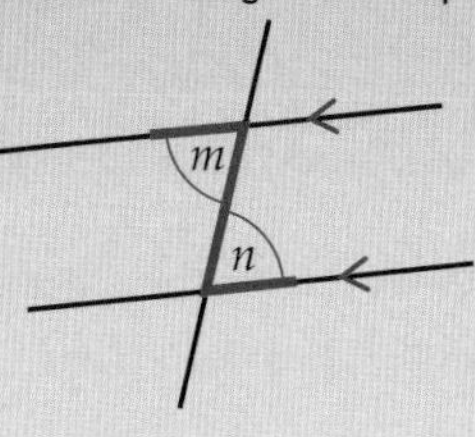

$m = n$

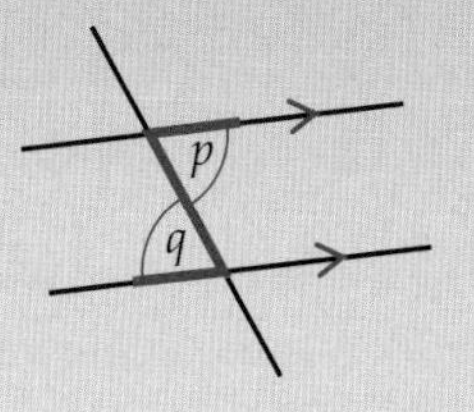

$p = q$

Bearings

E D C

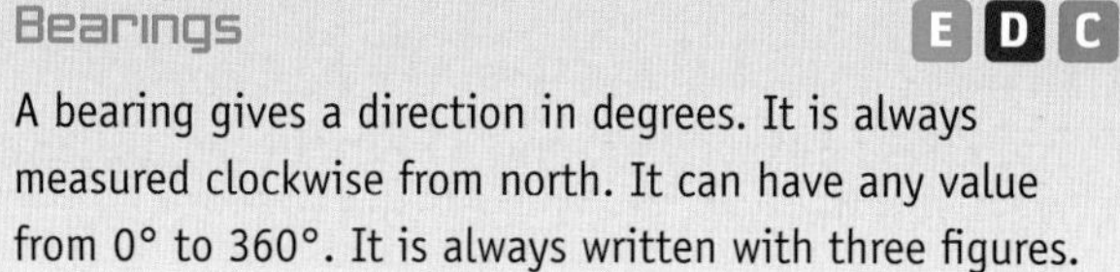

A bearing gives a direction in degrees. It is always measured clockwise from north. It can have any value from 0° to 360°. It is always written with three figures.

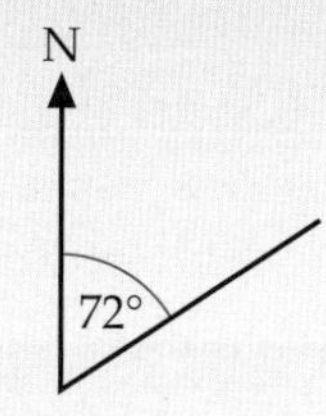

This shows a bearing of 072°.

21.1 Angle facts

E

1 Work out the sizes of the angles marked with letters.

a
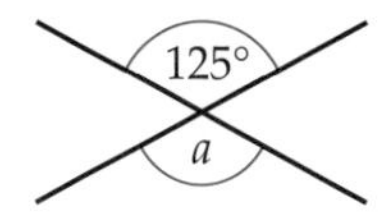

b
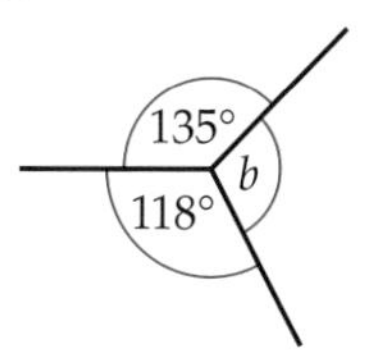

c
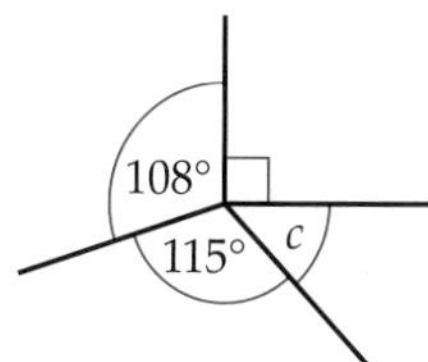

d
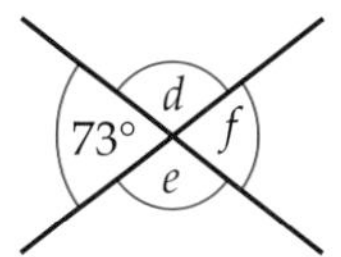

e
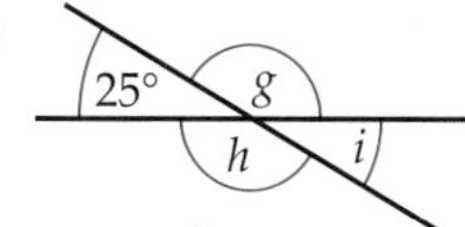

f
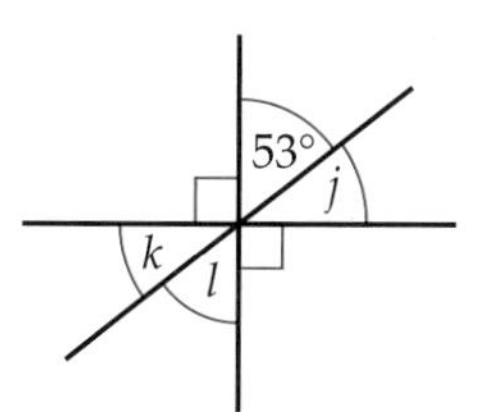

2 Work out the value of x in each diagram.

a

b
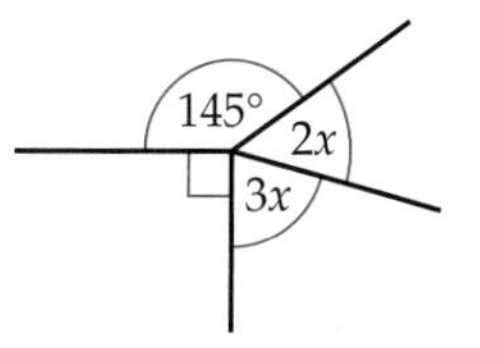

c
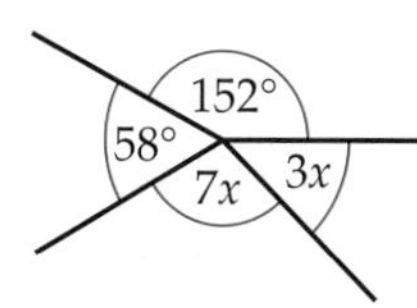

E

3 Work out the value of x and y in each diagram.

a

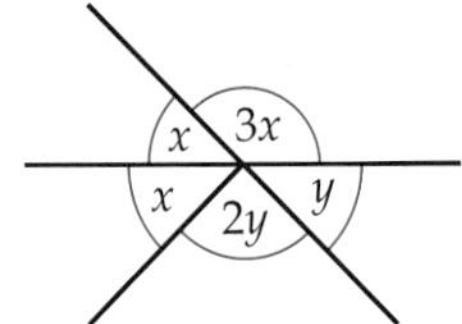

b

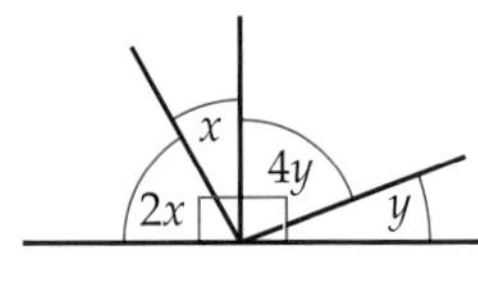

E

4 A wheel has five arms going out from the centre.
The angle between each pair of arms is the same.
Work out the angle between each pair of arms.

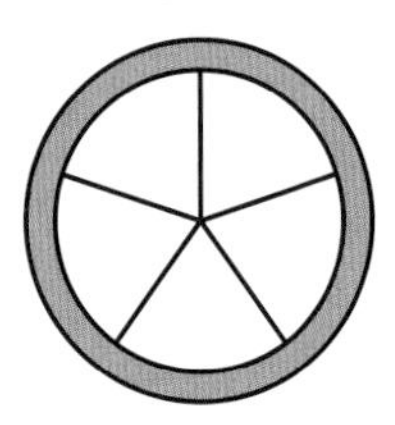

5 Sharon makes a design from identical kites.
The kites fit exactly around a point in two different ways.
Work out the sizes of angles x and y.

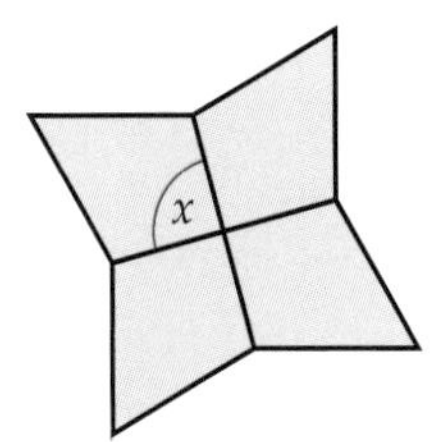

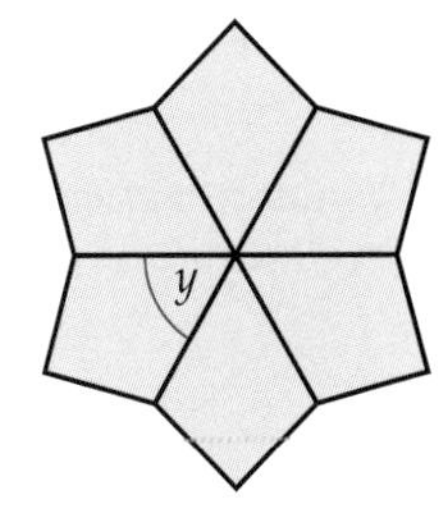

AO3

21.2 Angles in parallel lines

D

1 Work out the sizes of the angles marked with letters.
Give a reason for your answer each time.

> The reason could be alternate angles, corresponding angles or angles on a straight line.

a

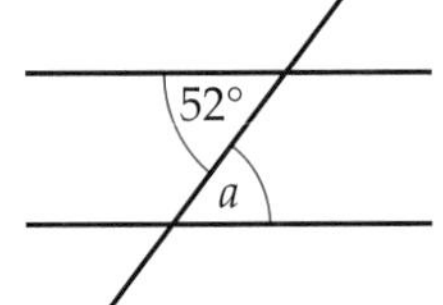

b

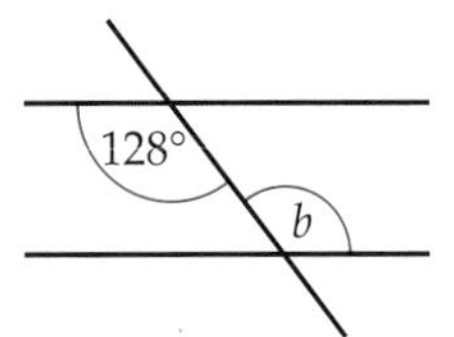

c

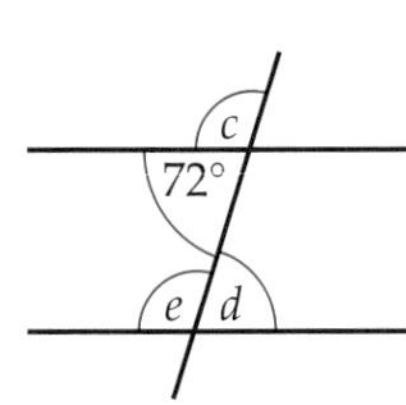

2 Work out the value of the angles marked with letters.
Give a reason for your answer each time.

> The reason could be alternate angles, corresponding angles, angles on a straight line or vertically opposite angles.

a

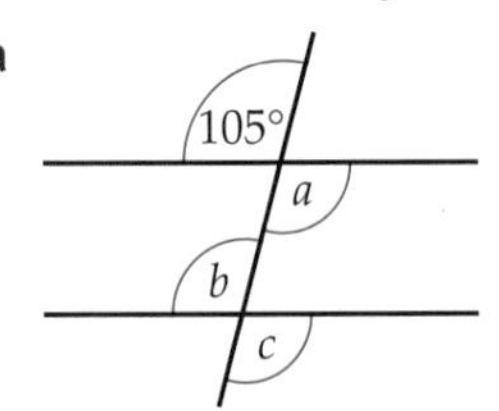

b

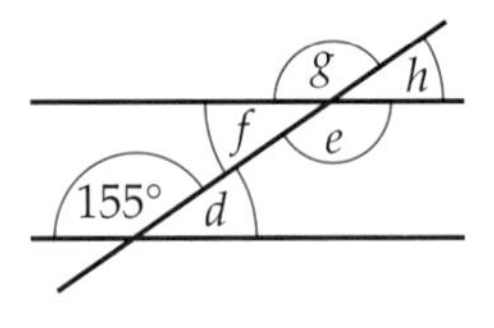

c

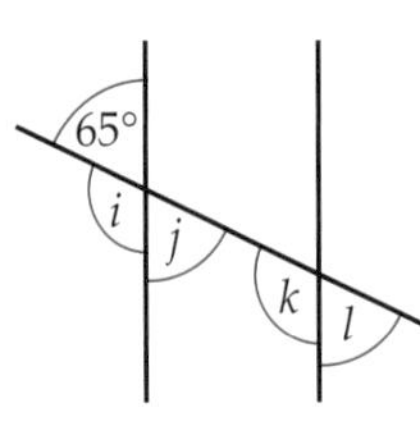

D

3 Show why angle $x = 55°$

> Copy the diagram and label any other angles you need.

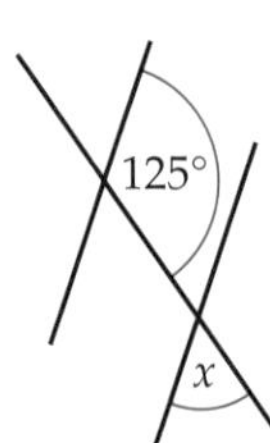

4 Show why angle $x = 79°$

> Remember that angles in a triangle add up to 180°.

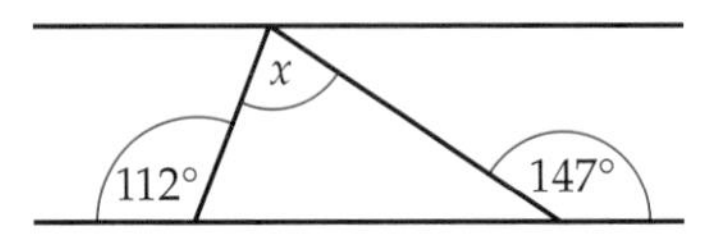

AO3

5 Sam has this isosceles trapezium.
In her isosceles trapezium, the top and bottom sides are parallel, and the other two sides are the same length. She colours one of the base angles green.
Sam draws this tessellation using her isosceles trapezium.

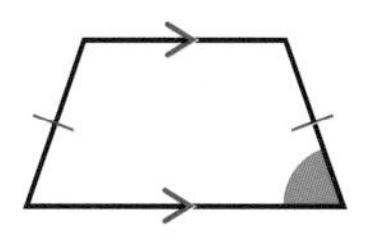

D

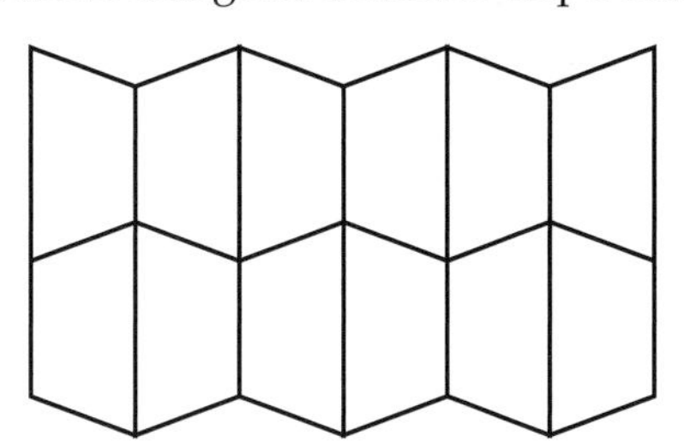

Sam colours in all of the angles that are the same size as the green angle.

a How many angles does she colour green?

Sam draws a diagonal line onto her trapezium and colours one of the top angles yellow.
The diagonal line bisects the green angle.
The green angle is 70°.

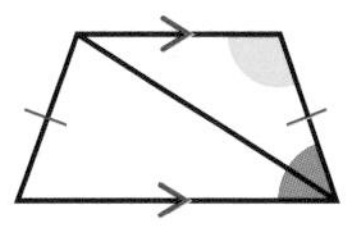

b Work out the size of the yellow angle.

AO3

21.3 Bearings

1 For each diagram measure the bearing of X from Y.

E

a

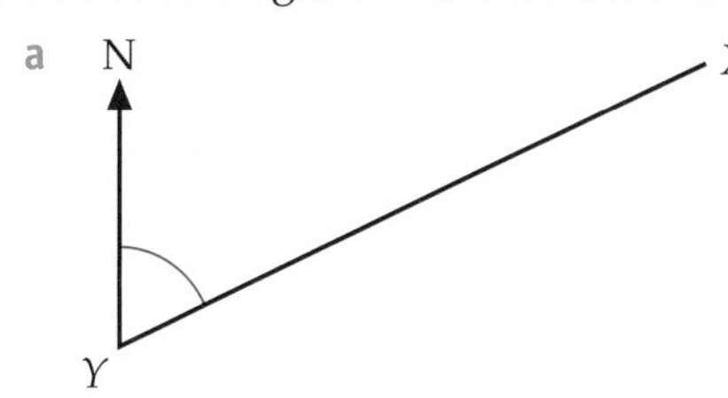

b

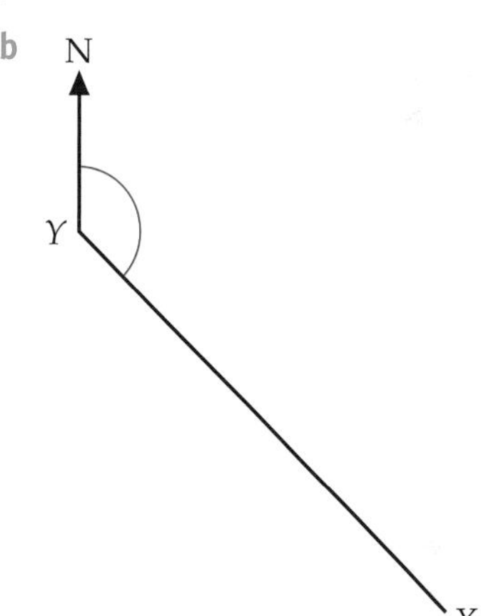

2 For each diagram measure the bearing of X from Y.

D

a

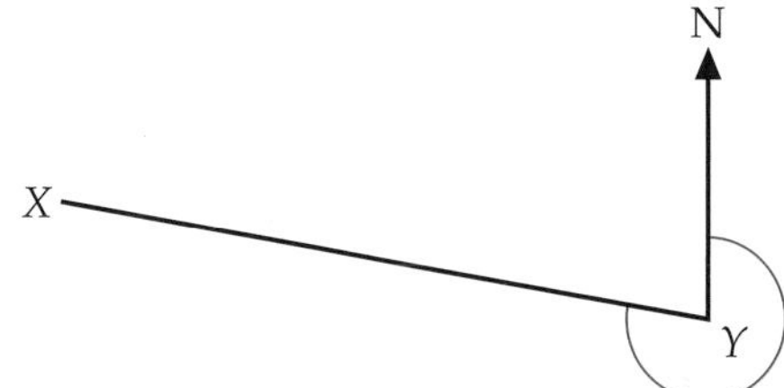

b

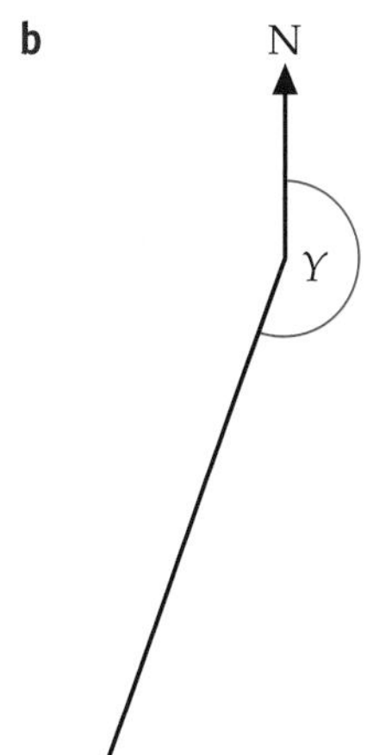

D

3 Draw accurate diagrams to show these bearings.

a 025° **b** 195° **c** 345°

D

4 Write down the missing numbers from each of these sentences.

a I am facing north. To go south-east I need to walk on a bearing of ☐°.

b I am facing north. To go south-west I need to walk on a bearing of ☐°.

c I am facing north. To go north-east I need to walk on a bearing of ☐°.

d I am facing north. To go north-west I need to walk on a bearing of ☐°.

Sketch a compass to help you.

5 The diagram shows the position of two people orienteering.

Dave ●

Barry ●

Dave runs on a bearing of 175° and Barry runs on a bearing of 265°.
Could Dave and Barry meet? Explain your answer.

AO3

C

AO3

6 A ship sails 5 km east. It then sails 7 km on a bearing of 165°.

a How far is the ship from its starting point?

b What bearing should the ship take to return to its starting point?

Draw a scale diagram.

C

7 For each diagram

i write down the bearing of A from B

ii work out the bearing of B from A.

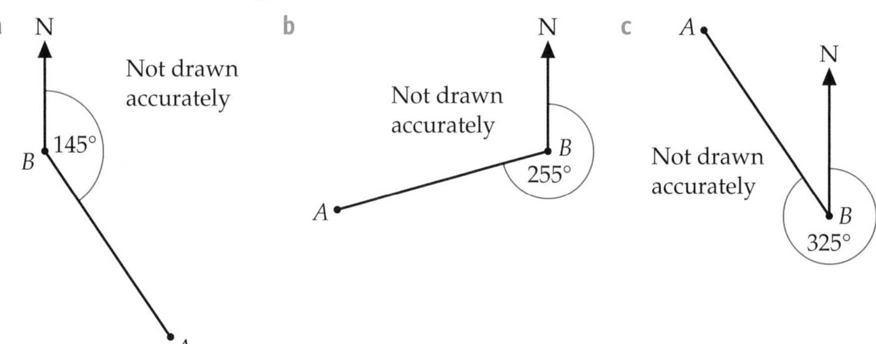

C

AO2

8 Carole rides her mountain bike from a gate to a fence on a bearing of 085°.

a Sketch Carole's route. Draw in a north line and label the angle.

b Carole rides from the fence back to the gate.
Use your diagram to work out her bearing for the return journey.

C

AO3

9 Paul is taking part in an orienteering competition.
The bearing to get from checkpoint C to checkpoint D is 125°.
Paul runs from checkpoint D to checkpoint C.
Work out the bearing on which Paul must run.

Links to:
Middle Student Book
Ch22, pp.357–367

Key Points

Time and timetables E

The 12-hour clock uses am for the morning and pm for the afternoon.

The 24-hour clock counts a whole day from 00 00 to 23 59.

Converting units E

You can convert between different metric units by multiplying or dividing by 10, 100 or 1000.

You can also measure lengths, capacities and masses using imperial units.

You need to know the approximate conversions between metric and imperial units.

Maps and scale drawings E

A scale drawing has the same proportions as the object it represents; a map is a type of scale drawing.

The scale tells you the relationship between lengths on the drawing or map and lengths in real life. Scales for maps are usually given as ratios.

Accuracy of measurements C

Measurements that are given to the nearest whole number might be up to half a unit larger or smaller than the given value.

The smallest possible actual value for a measurement is called the lower bound or minimum value.

The largest possible actual value for a measurement is called the upper bound or maximum value.

22.1 Time and timetables

1 Carole works part time.
This time sheet show the hours she works one week.

Day	Start time	Finish time	Time worked
Monday	09 30	12 40	3 h 10 min
Tuesday	10 30	12 55	
Wednesday	14 00	16 15	
Thursday	14 15	17 30	
Friday	08 30	13 55	

a Copy and complete Carole's time sheet.

b Work out the total time that Carole worked during this week.

c Carole is paid £12 per hour.
How much did Carole earn this week?

2 The coach journey from Liverpool to Carlisle should take 3 hours and 45 minutes.
Andy's coach leaves Liverpool at 12:27 pm.

a What time should Andy arrive in Carlisle?

b His coach is delayed by 25 minutes.
What is his new arrival time?

E

3 This is part of a bus timetable in Manchester.

MONDAYS TO FRIDAYS Except Bank Holidays														
Bramhall Village, Moss lane	05 58	06 28	06 51	07 08	07 38	……	08 30	……	08 55	09 26	09 55	10 25	10 55	11 25
Grove Lane, Smithy Hotel	06 02	06 32	06 56	07 16	07 46	08 20	08 37	……	09 00	09 30	10 00	10 30	11 00	11 30
Cheadle Hulme, Rail Station	06 07	06 37	07 03	07 25	07 55	08 29	08 46	0854	09 06	09 36	10 05	10 36	11 06	11 36
Cheadle & Marple College	06 10	06 40	07 07	07 31	08 01	08 35	08 52	0859	09 10	09 40	10 10	10 40	11 10	11 40
Cheadle, Post Office	06 14	06 44	07 12	07 38	08 08	08 42	08 59	0908	09 15	09 45	10 15	10 45	11 15	11 45
East Didsbury, Parrs Wood	06 18	06 48	07 17	07 45	08 15	08 49	09 04	0911	09 20	09 50	10 20	10 50	11 20	11 50
Didsbury Village, Co-op	06 21	06 51	07 21	07 50	08 20	08 54	09 09	0916	09 24	09 54	10 24	10 54	11 24	11 54
Fallowfield, Friendship	06 26	06 56	07 26	07 55	08 25	08 59	09 14	0921	09 29	09 59	10 29	50 59	11 29	11 59

a At what time does the 08 42 bus from Cheadle Post Office arrive at Fallowfield, Friendship?

b How long is the journey from the Rail Station at Cheadle Hulme, leaving at 08 54, to the Post Office at Cheadle?

c The 09 26 bus from Moss Lane in Bramhall Village is delayed by 7 minutes. What time will it arrive at the Co-op in Didsbury Village?

d Essien needs to be at Parrs Wood by 9:00 am. What is the latest bus he can catch from Grove Lane?

e Do you think it's further from Smithy Hotel, Grove Lane to Cheadle Post Office, than from Cheadle Post Office to the Co-op, Didsbury Village? Give a reason for your answer.

22.2 Converting between metric and imperial units

E

1 Copy and complete these conversions.

a 5 miles ≈ ☐ km

b 25 miles ≈ ☐ km

c ☐ miles ≈ 12 km

d 1 inch ≈ ☐ cm

e 7 inches ≈ ☐ cm

f ☐ inches ≈ 18 cm

g 1 kg ≈ ☐ pounds

h 12 kg ≈ ☐ pounds

i ☐ kg ≈ 33 pounds

j 1 gallon ≈ ☐ litres

k 8 gallon ≈ ☐ litres

l ☐ gallon ≈ 27 litres

2 Elisa is driving in Spain.
She sees this sign.
How many miles is Lisa from Barcelona?

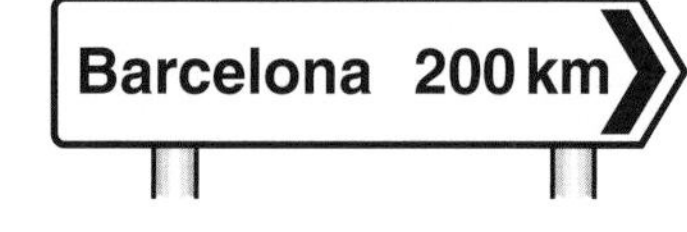

3 Greg buys a new lawnmower. It has an 18-inch blade.
How many centimetres is this?

4 Orlando puts 45 litres of petrol in his car.
How many gallons is this?

E
AO2

5 Hannah is going on holiday. The maximum her luggage can weigh is 22 kg.
According to her bathroom scales her luggage weighs $3\frac{1}{2}$ stone.
There are 14 pounds in a stone.
Is Hannah's luggage too heavy? Explain your answer.

E

1 A scale drawing uses a scale of 1 cm to represent 20 m.

a Work out the length on the drawing of each of these real-life lengths.
  i 80 m   ii 110 m

b Work out the length in real-life of each of these lengths on the drawing.
  i 8 cm   ii 2.7 cm

2 Asher has made a scale drawing of her kitchen on centimetre squared paper.

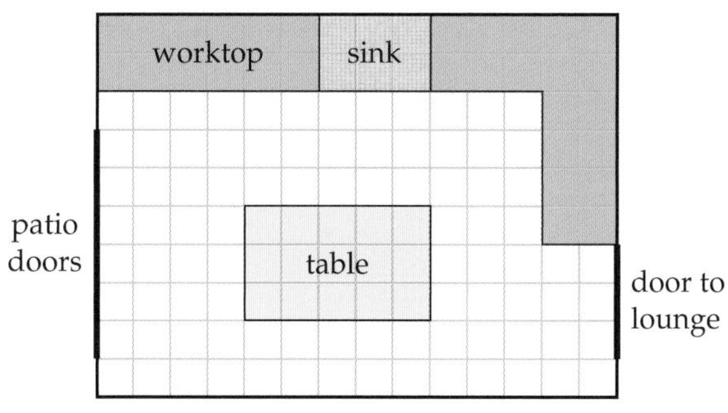

Scale: 1 cm represents 30 cm

a How wide are the patio doors?

b Kitchen cupboards are 60 cm wide.
How many cupboards can Asher fit under the worktop between the sink and the left wall?

c Asher works out how many people she can seat at her table.
She estimates that one person needs at least a 50 cm wide piece of the table.
Asher says, 'I think I can seat 10 people comfortably at the table.'
Is Asher correct? Explain your answer.

3 Andy is competing in a sponsored cycle ride.
The diagram shows a map of the cycle route.
The scale of the map is 1 : 500 000.

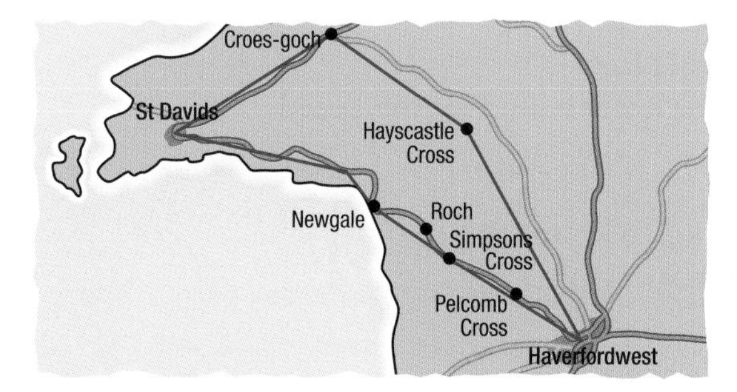

a Convert 500 000 cm into kilometres.

b What distance would be represented by 5 cm on the map?

c What length on the map would be represented by a distance of 1 km?

d Measure the distance from Croes-goch to St Davids.

e Work out the real-life distance from Croes-goch to St Davids.

f The route of the cycle ride is shown in red on the map.
Work out an approximate length of the whole cycle ride.

1 John weighs 72 kg to the nearest kilogram.
Write down
- a the upper bound for his weight
- b the lower bound for his weight.

2 The length of a garden is 18 m to the nearest metre.
Write down
- a the maximum value for the length
- b the minimum value for the length.

3 The lengths of this triangle are measured to the nearest centimetre.
Work out
- a the minimum values for the dimensions of the triangle
- b the minimum value for the area of the triangle.

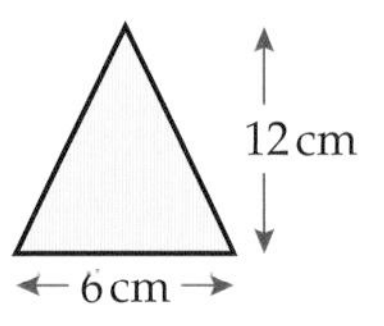

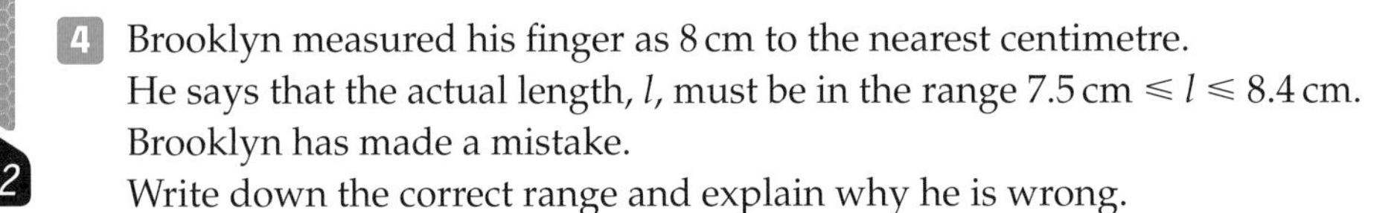

4 Brooklyn measured his finger as 8 cm to the nearest centimetre.
He says that the actual length, l, must be in the range $7.5\text{ cm} \leqslant l \leqslant 8.4\text{ cm}$.
Brooklyn has made a mistake.
Write down the correct range and explain why he is wrong.

AO2

5 The distance between two towns is given as 260 km to the nearest 10 km.
Work out
- a the maximum distance between the two towns
- b the minimum distance between the two towns.

6 Pendine Sands is an 11.2 km long beach in Carmarthenshire.
The first person to use Pendine Sands for a world land speed record attempt was Malcolm Campbell. In 1924 he set a record of 235.22 km/h in his car Blue Bird.
- a 11.2 km is correct to one decimal place.
  Write down the upper bound and the lower bound for this distance.
- b 235.22 km/h is correct to two decimal places.
  Write down the upper bound and the lower bound for this speed.

7 Five people want to get in a lift.
The five people weigh 92 kg, 75 kg, 64 kg, 88 kg and 95 kg.
All their weights are measured to the nearest kg.
- a Work out
  - i the total maximum weight of the people
  - ii the total minimum weight of the people.
- b The lift is allowed to carry a total weight of 420 kg measured to the nearest 10 kg.
  Work out the maximum weight the lift can take.
- c Is it safe for all five people get in the lift? Give a reason for your answer.

AO2

23 Triangles and construction

Links to:
Middle Student Book
Ch23, pp.368–377

Key Points

Types of triangles E

In a right-angled triangle, one angle is 90°.

In a scalene triangle, all three sides and angles are different.

In an isosceles triangle, two angles and two sides are the same.

In an equilateral triangle, all three sides and angles are the same.

Triangle properties E

The sum of the interior angles of a triangle is 180°.

The exterior angle of a triangle is equal to the sum of the two opposite interior angles.

Constructing triangles E D

To accurately construct a triangle given all three sides, use a ruler and compasses.

To accurately construct a triangle given SAS (side angle side) or ASA (angle side angle), use a ruler and a protractor.

Leave in all your construction lines.

Congruent triangles C

The four conditions for congruent triangles are:

- SSS (side side side)
- SAS (side angle side)
- ASA (angle side angle) or SAA (side angle angle)
- RHS (right angle hypotenuse side).

23.1 Interior and exterior angles

E

1 Work out the size of each angle marked with a letter.

a

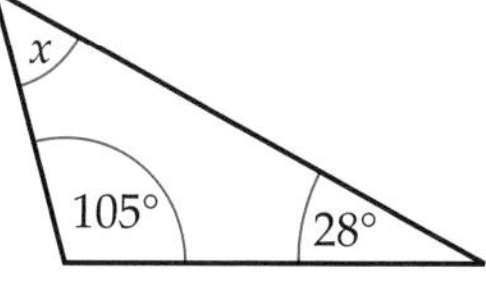

b

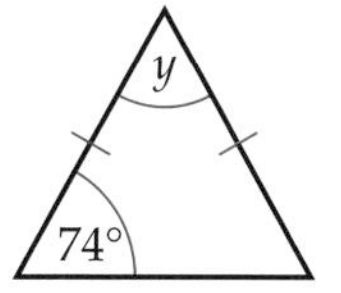

c

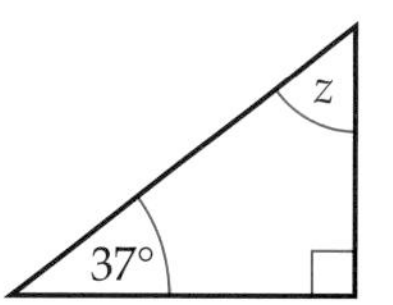

E

2 Work out the size of each angle marked with a letter.

a

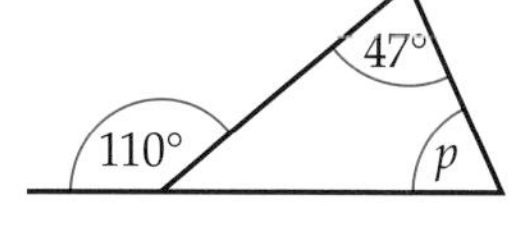

b

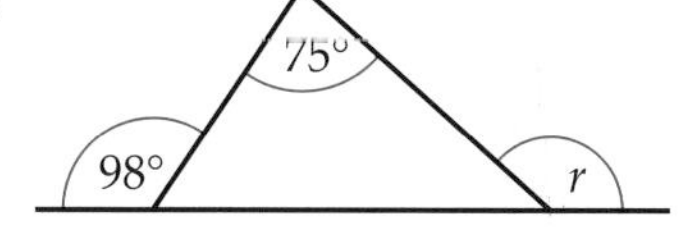

c

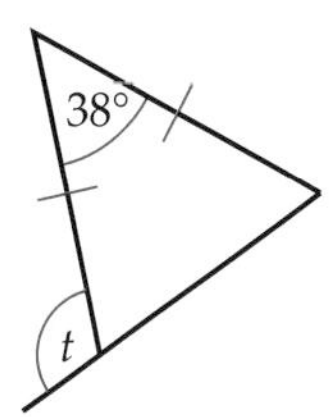

d

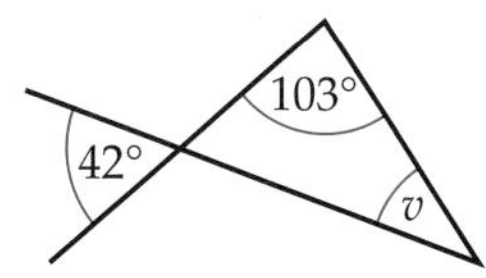

e

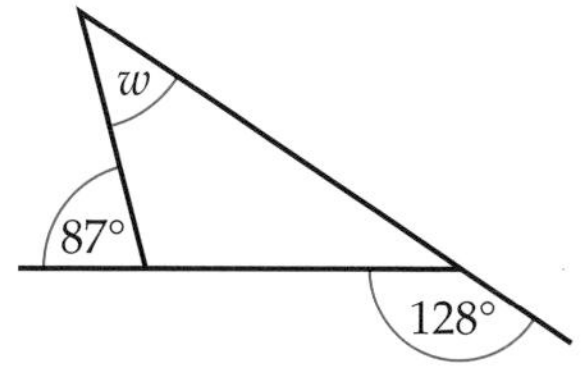

AO2

D

3 Work out the size of angle x in each triangle.

a

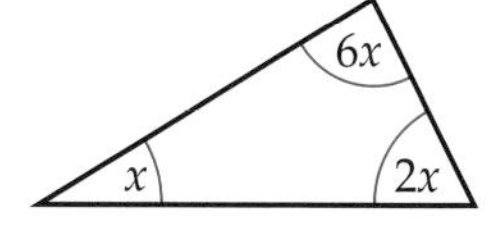

b

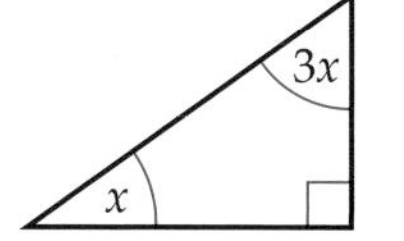

c

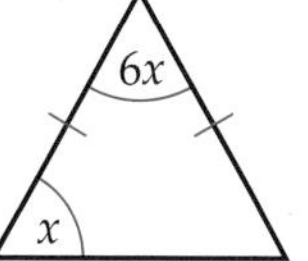

d

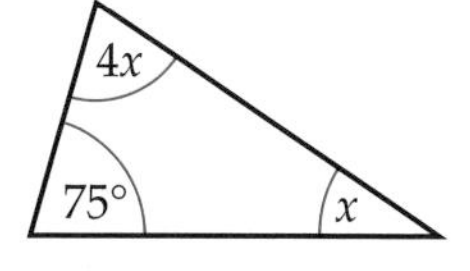

e

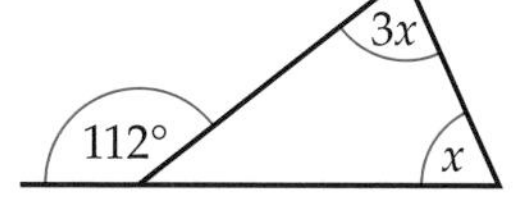

AO2

D

A02

4 Work out the value of a in this diagram.

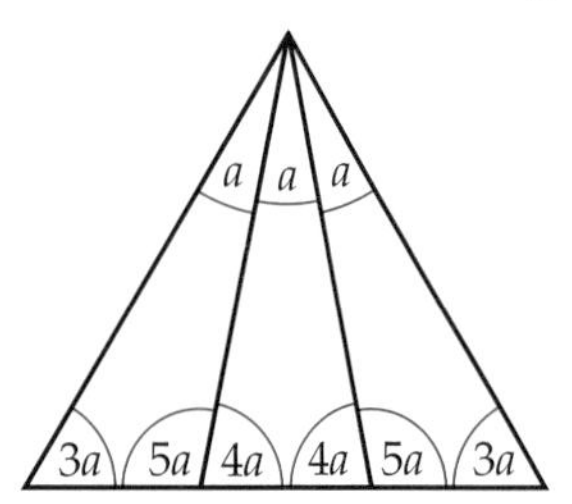

23.2 Constructions

E

1 a Construct triangle ABC with the following measurements.
$AB = 9.5$ cm, $AC = 4$ cm and $BC = 7.5$ cm.

b Name the type of triangle you have drawn.

c Measure and write down the size of angle BAC.

d Measure and write down the size of angle ACB.

2 Draw an accurate copy of this shape.

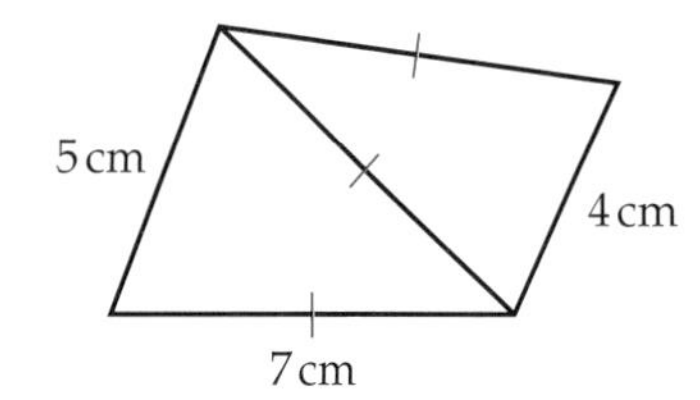

D

3 **a** Draw an accurate copy of each of these triangles.

i

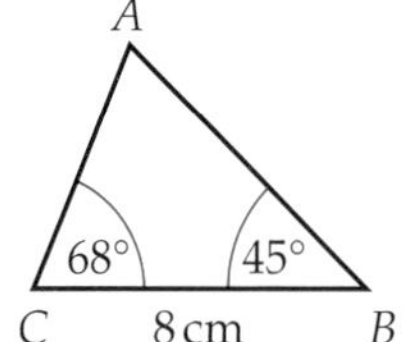

ii

b Write down the length AB for each of the triangles you drew in part **a**.

c Measure and write down the size of angle BAC in each of the triangles you drew in part **a**.

D

A02

4 **a** Draw an accurate copy of this triangle.

b Measure and write down the size of angle BAC.

c Measure and write down the length AC.

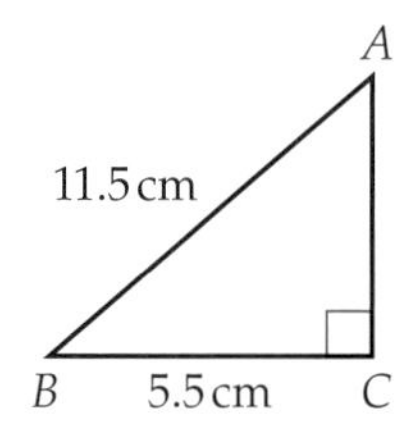

D

A03

5 An architect is finishing the design of a children's slide.
This is the sketch she has made.

The architect says that the ladder is about 2.3 m long.
Is the architect correct?

Draw a scale drawing of the slide.

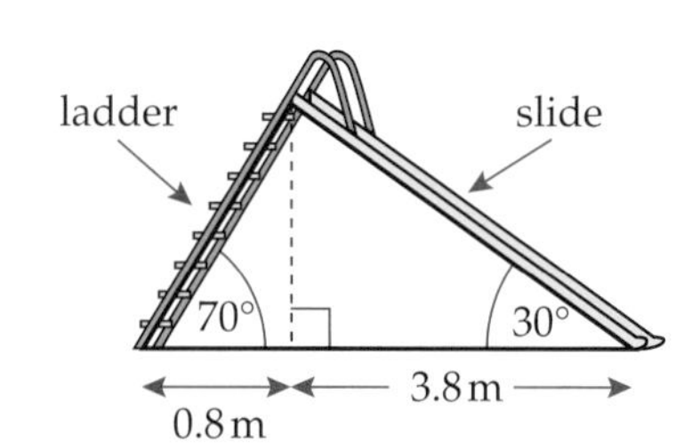

C

1 Write down the four conditions that can show that two triangles are congruent.

2 Triangles A and B are congruent.
Write down the condition which shows that the triangles are congruent.

A

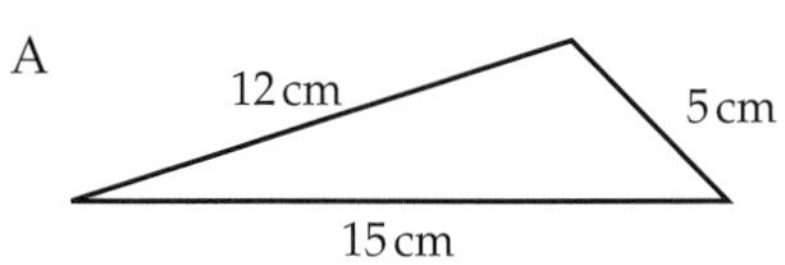

B

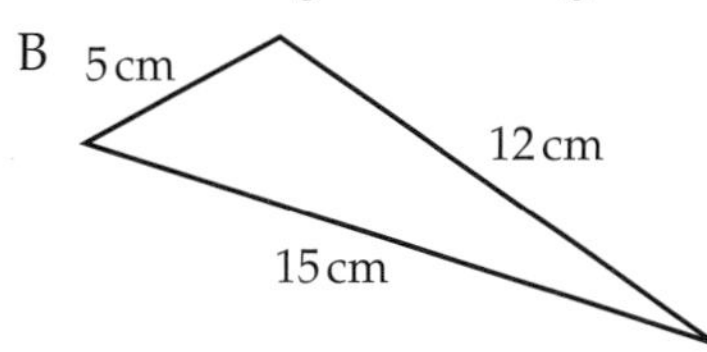

3 Triangles C and D are congruent.

C

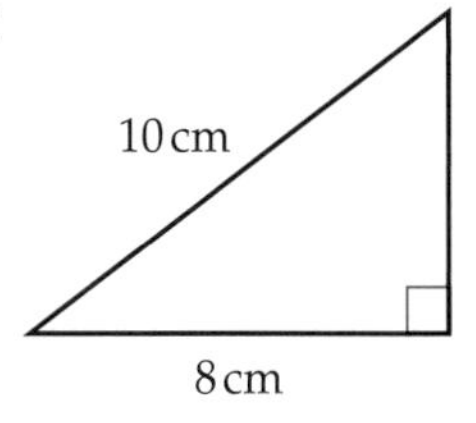

D

10 cm
8 cm

Write down the condition which shows that the triangles are congruent.

4 State whether or not the following pairs of triangles are congruent.
If they are congruent, write down which one of the four conditions is satisfied.

a

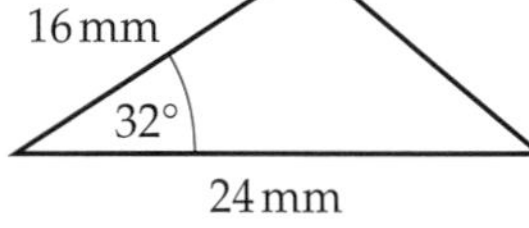

16 mm
32°
24 mm

b

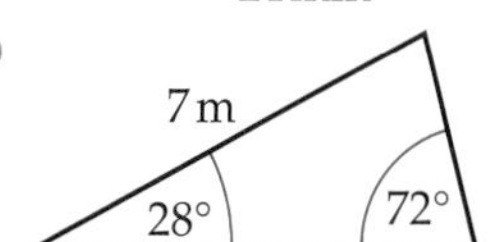

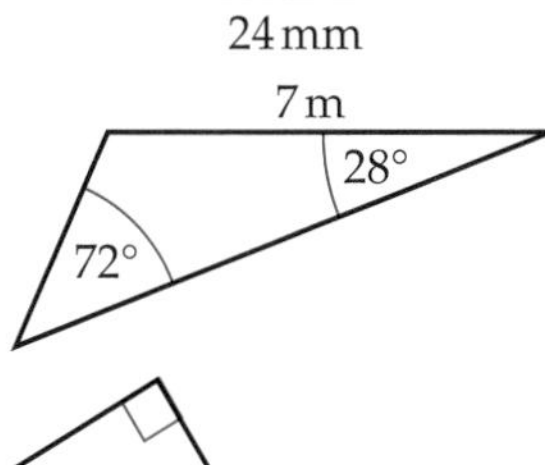

c

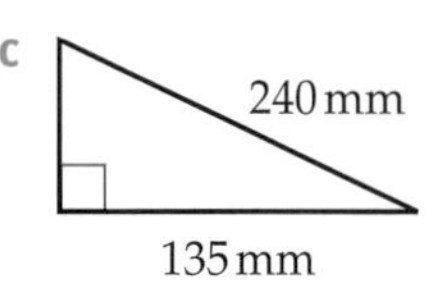

13.5 cm
24 cm

5 Alison draws a triangle congruent to this one.

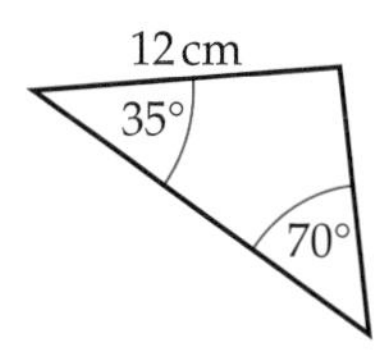

This is what she draws.

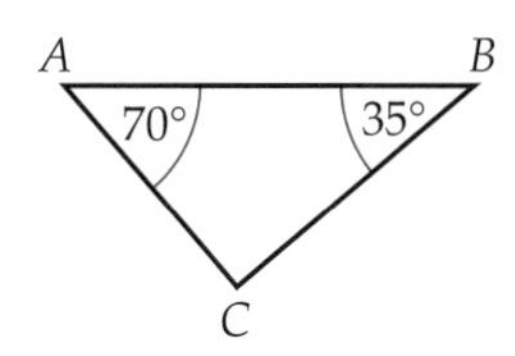

Which side, *AB*, *BC* or *AC*, must Alison label 12 cm?

Links to:
Middle Student Book
Ch24, pp.378–389

Key Points

Writing your own equations D C B

You can solve problems by writing and solving equations.

Writing your own formulae D C

You can use letters or words to write your own formulae.

You can substitute values into formulae to solve problems.

Changing the subject of a formula C B

The subject of a formula is the letter on its own.

You can use the rules of algebra to change the subject of the formula.

Changing the subject of a formula is like solving an equation. You need to get a letter on its own on one side of the formula.

Proof C

A proof is a mathematical argument.

When you prove something you need to explain each step of your working.

Trial and improvement C

Some equations cannot be solved using algebra.

Use trial and improvement to find solutions to these equations. Substitute a value of x into the equation and see how close it is to the value you want. Try again with a different value. The more values you try the closer you can get to the solution.

24.1 Equations and formulae

C

1 Work out the value of the letter in each of these diagrams.

a a, $a + 40°$, $a + 20°$

b $x + 5°$, $2x + 10°$, $x + 5°$

c $2c + 50°$, $3c + 20°$

D A02

2 The regular pentagon and regular hexagon have the same perimeter.
The length of one side of the pentagon is 9 cm.
Work out the length of one side of one side of the hexagon.

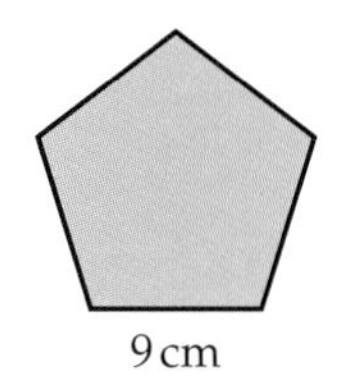

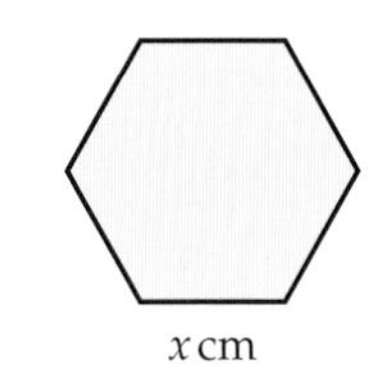

D A03

3 This square and regular octagon have the same perimeter.
The area of the square is 81 cm^2.
Work out the length of one side of one side of the octagon.

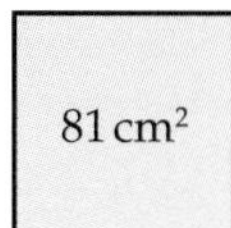

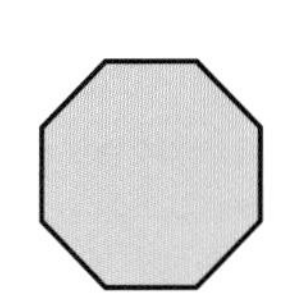

B A02

4 The square of a number, plus 70, is equal to 19 times the number.

a Write a quadratic equation to show this information.

b Solve your equation to find two possible values of the number.

C A02

5 The length of a rectangle is 10 cm more than its width, w.

a Write a formula for the area of the rectangle, A.

b Work out A when $w = 4$ cm.

c Work out w when $A = 24$ cm^2.

6 The formula for the volume of a cone is

$$\text{volume} = \tfrac{1}{3} \times \text{base area} \times \text{height}$$

a Rearrange this formula to make 'base area' the subject.

b Use your rearranged formula to find the base area of this cone

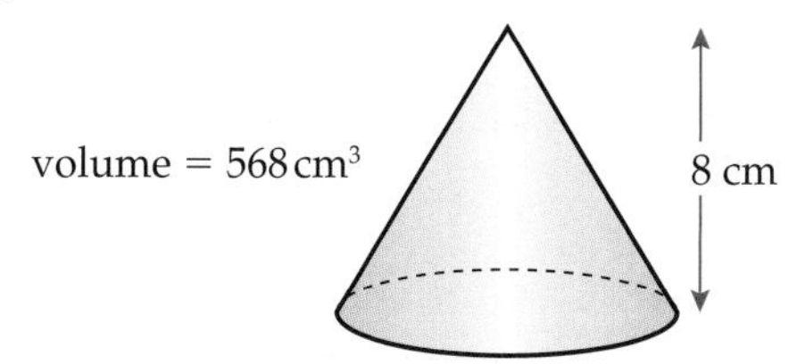

C

AO2

24.2 Proof

1 The diagram shows a rhombus.
Prove that the opposite angles in a rhombus are the same size.

Start by drawing in one of the diagonals of the rhombus.

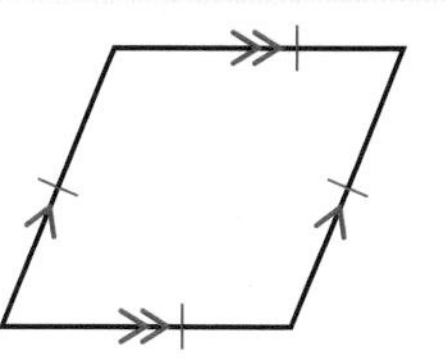

2 Prove that the sum of the interior angles in a pentagon is 540°.

You can use the fact that the angles in a triangle add up to 180°.

3 Prove that each exterior angle of a regular pentagon is 72°.

C

AO3

24.3 Trial and improvement

1 Dom is using trial and improvement to solve the equation $x^3 + 7x = 200$.
His first two trials are shown in the table.

x	$x^3 + 7x$	Comment
4	92	too low
6	258	too high

Copy the table, and use it to help you solve the equation.
Give your answer correct to 1 d.p.

C

2 The area of this rectangle is 400 cm².

$(x^2 + 2)$ cm

x cm

a Write an equation showing this information.

b Use trial and improvement to work out the value of x.
Give your answer correct to 1 d.p.

C

AO2

3 The capacity of a water tank is given by the formula

$$V = d^3 - \frac{d}{0.8}$$

where V is the volume in cubic metres and d is the depth of the tank.
Use trial and improvement to find the value of d when $V = 150\,\text{m}^3$.
Give your answer correct to 1 d.p.

C

AO3

Quadrilaterals and other polygons

Links to:
Middle Student Book
Ch25, pp.390–403

Key Points

Quadrilateral and polygon properties

E D C

A quadrilateral is a flat shape bounded by four straight lines.

The angle sum of a quadrilateral is 360°.

The sum of the exterior angles of any polygon is 360°.

Exterior angle of a regular polygon $= \frac{360°}{\text{number of sides}}$

Number of sides of a regular polygon $= \frac{360°}{\text{exterior angle}}$

The sum of the interior angles of any polygon = (number of sides − 2) × 180°.

Interior angle of a regular polygon $= 180° - \frac{360°}{\text{number of sides}}$

Number of sides of a regular polygon $= \frac{360°}{180° - \text{interior angle}}$

Plotting geometric shapes

E

You can draw geometric shapes on a coordinate grid by plotting the coordinates of the vertices.

25.1 Quadrilaterals and algebra

E

1 Calculate the sizes of the angles marked with letters.

a

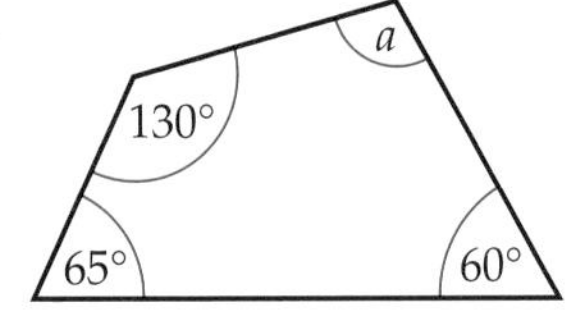

b

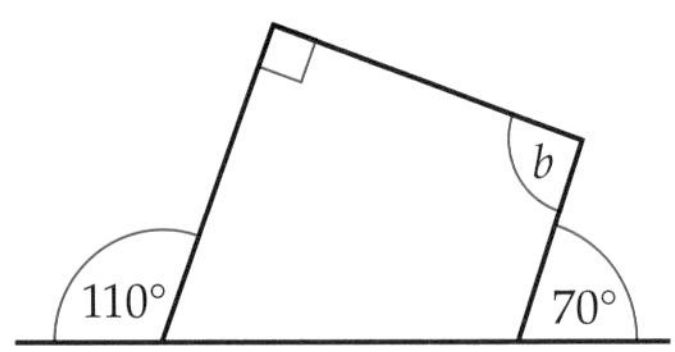

E A02

2 In this quadrilateral the smallest angle is 20°.
The opposite angle is three times as large.
Another angle is three times as large as twice the smallest angle.
What is the size of the remaining angle?

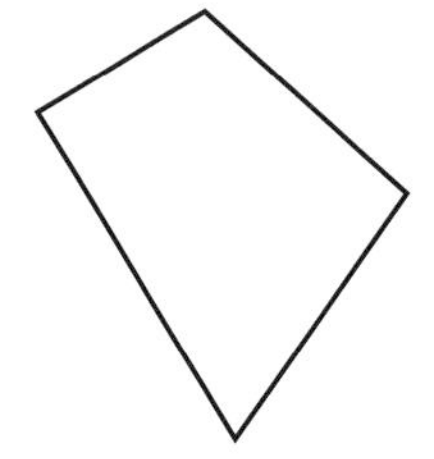

D

3 For each of these quadrilaterals

i form an equation in x

ii solve the equation to find the value of x.

a

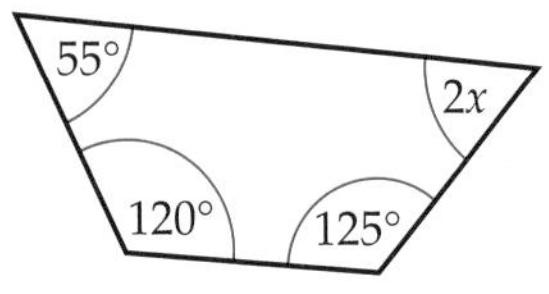

b

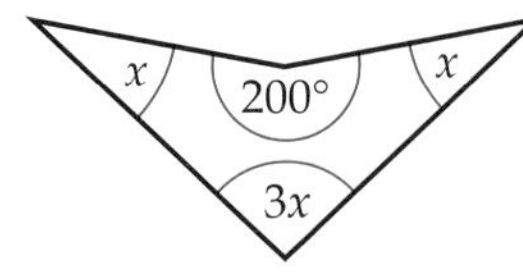

D

4 A quadrilateral has angles with values x, $3x$, $5x$ and $7x$.
Calculate the value of the largest angle.

A02

5 Jaycee says, 'The largest angle in this quadrilateral is the same as the sum of the two smallest angles.'
Is Jaycee correct? Show working to support your answer.

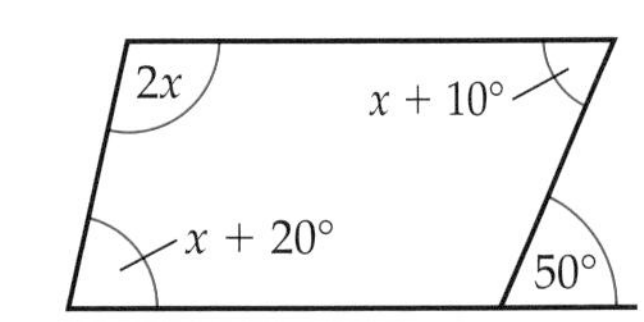

25.2 Special quadrilaterals

E

1 Show how you can join together three isosceles triangles to make a trapezium.

AO2

2 Show how you can join together four isosceles triangles to make a parallelogram.

D

3 Work out the sizes of the angles marked with letters.

a

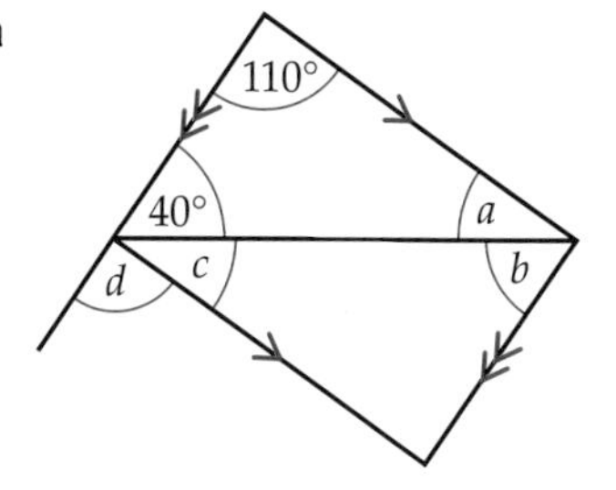

b

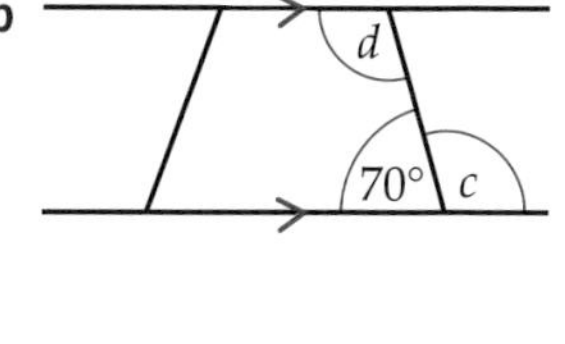

c

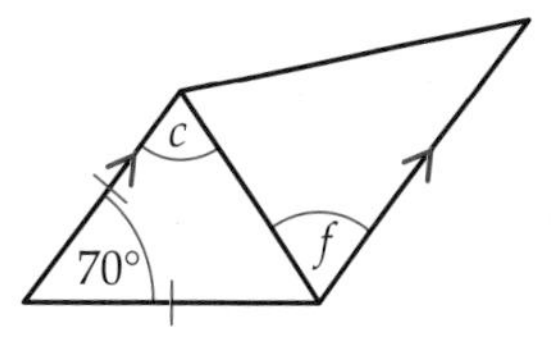

D

4 Work out the sizes of the angles marked with letters.

a

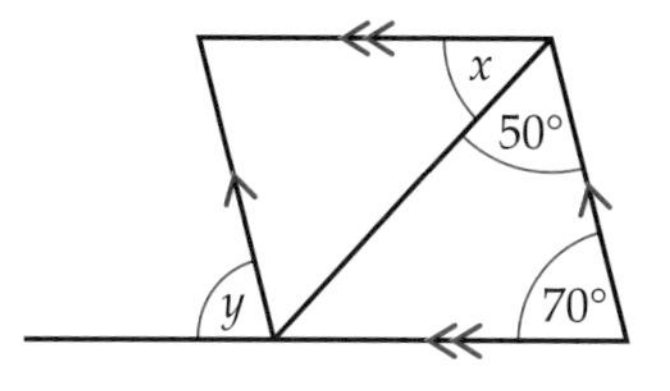

b

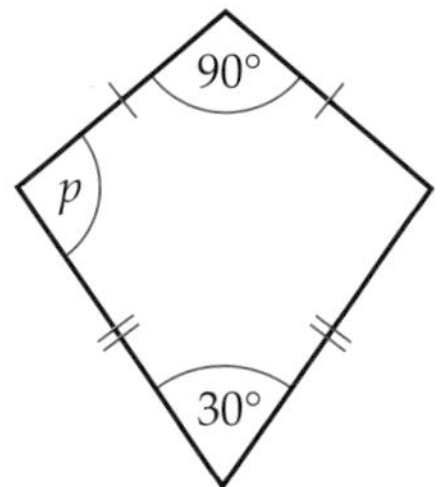

c

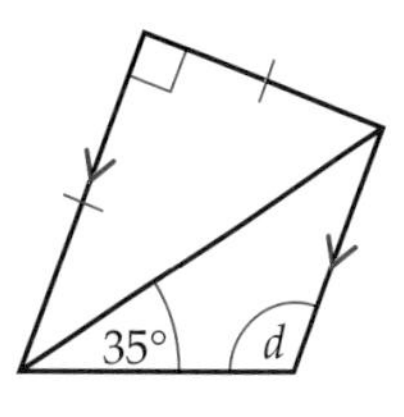

AO2

5 Juan says, 'In this chevron angle x is 140°.'
Matilda says, 'I think angle x is 80°.'
Who is correct? Show working to support your answer.

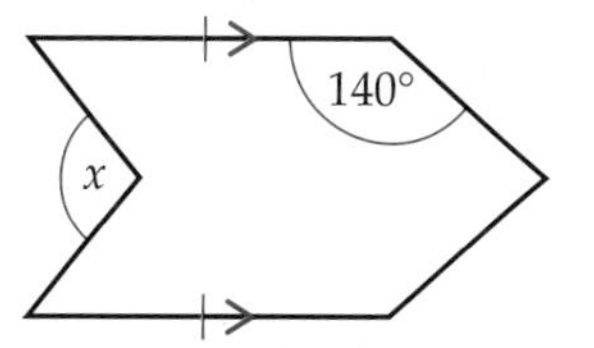

25.3 Polygons

D

1 Work out the size of the lettered angle in each of these polygons.

a

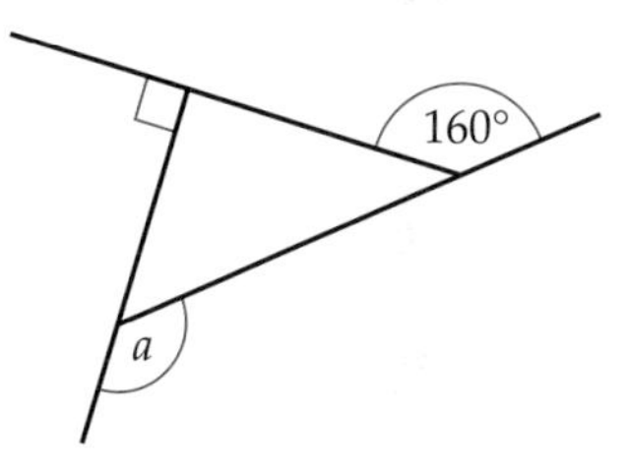

b

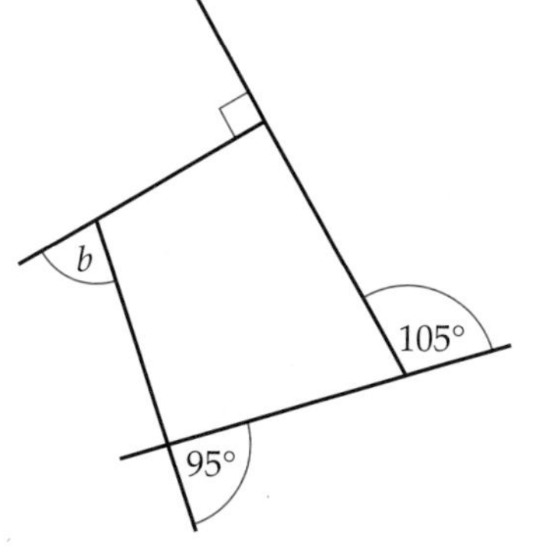

c

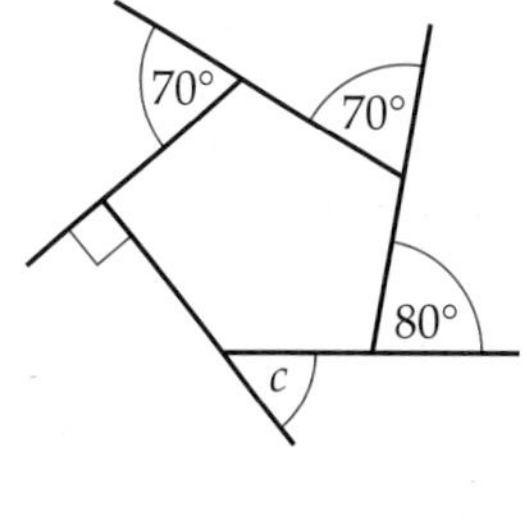

C

2 Work out the size of the exterior angle of a regular polygon with 18 sides.

C

3 Explain why it is not possible for the exterior angle of a regular polygon to be 14°.

AO2

D

4 Work out the size of angle x in each of these polygons.

a

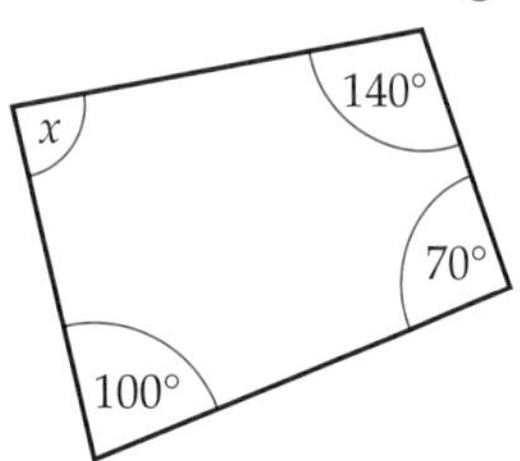

b

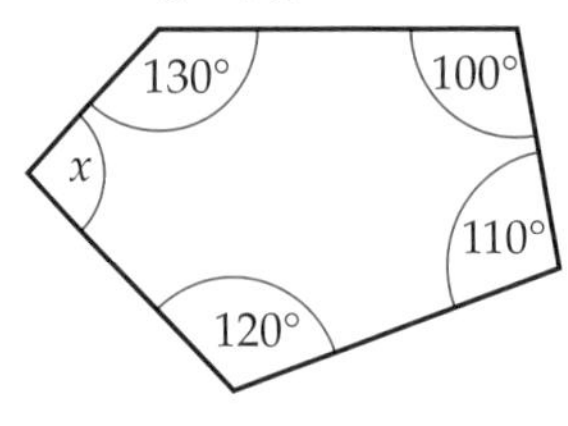

c

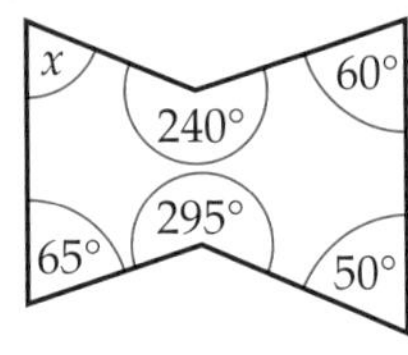

C

5 Work out the size of the interior angle of a regular polygon with 12 sides.

6 How many sides does a regular polygon have if the interior angle is 165°?

C

7 Explain why it is not possible for the interior angle of a regular polygon to be 130°.

8 A carpenter is making a large wooden table-top.
The table-top is in the shape of a regular decagon (10-sided shape).
The diagram shows a plan view of the table-top.
At what angle, marked as x on the diagram, must the carpenter cut the wood for the table-top?

AO2

25.4 Coordinates

E

1 Copy the diagram.

a Write down the coordinates of A.

b Write down the coordinates of B.

c Write down the coordinates of C.

d Mark the mid-point of AC with a cross and label it E.

e Write down the coordinates of E.

f Plot the point D $(-1, 3)$ on your diagram.

g Join A to B to C to D to A.

h What is the mathematical name of the quadrilateral $ABCD$?

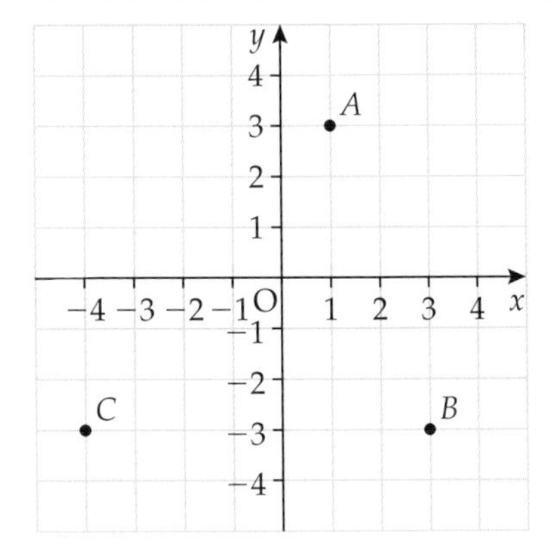

E

2 Hamish plots the three points shown on the grid.
Hamish says, 'If I plot another point at $(-1, 2)$ I'll get a kite.'
Is Hamish correct? Explain your answer.

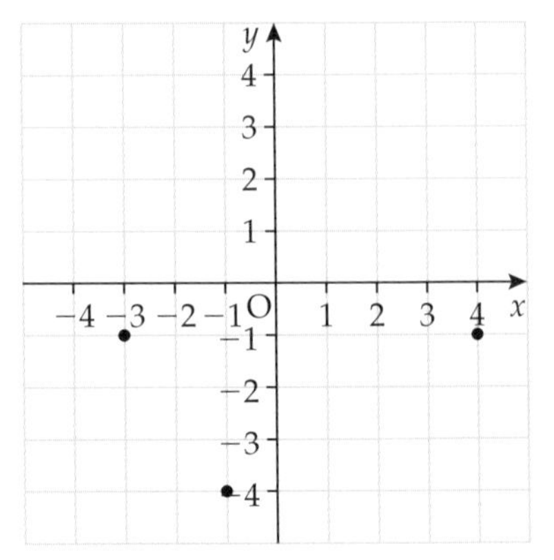

AO2

Links to:
Middle Student Book
Ch26, pp.404–415

Key Points

Perimeter and areas of simple shapes E D

The perimeter of a shape is the sum of the lengths of all its sides.

The area of a shape is the amount of space inside it.

Rectangle

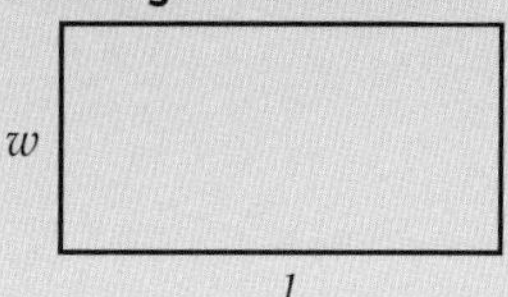

Perimeter $= 2l + 2w$

Area $=$ length $\times$ width

$= l \times w$

Parallelogram

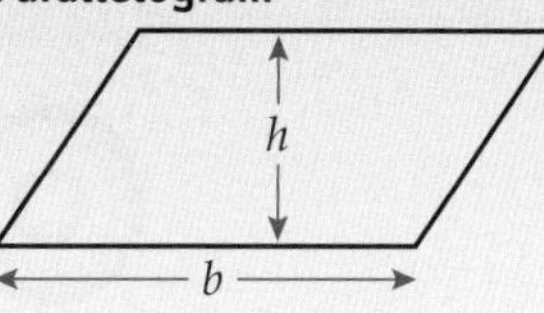

Area $=$ base $\times$ perpendicular height

$= b \times h$

Triangle

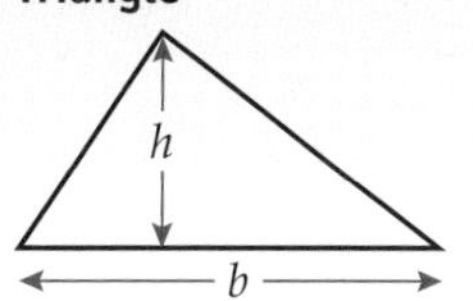

Area $= \frac{1}{2} \times$ base $\times$ perpendicular height

$= \frac{1}{2} \times b \times h$

Trapezium

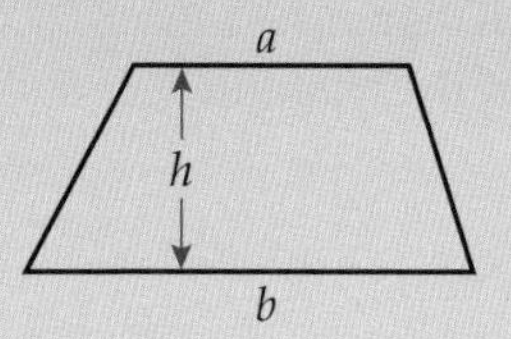

Area of trapezium $= \frac{1}{2} \times$ (sum of parallel sides) $\times$ perpendicular height

$= \frac{1}{2} \times (a + b) \times h$

Areas of compound shapes E D

To find the area of a compound shape you split it into simple shapes. Then you use the formula for the area of each shape separately.

Prisms

A prism is a 3-D shape whose cross-section is the same all through its length.

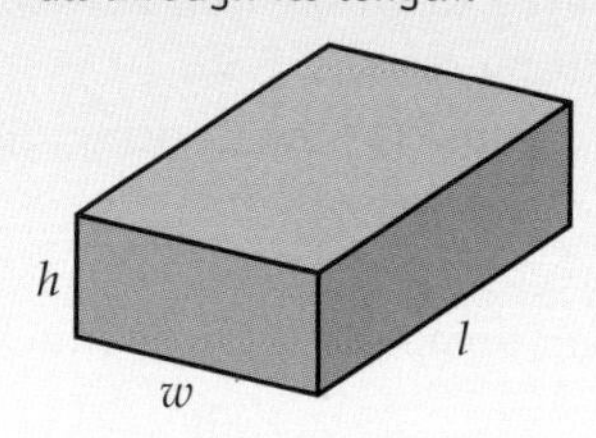

Volume of cuboid $=$ length $\times$ width $\times$ height

$= l \times w \times h$

Volume of prism $=$ area of cross-section $\times$ length

The surface area of a prism is the sum of the areas of all its surfaces.

26.1 Perimeter and area of simple shapes

E

1 Calculate the perimeters and areas of these rectangles.

a 9 cm, 4 cm

b 2 cm, 15 cm

D

2 Calculate the perimeters and areas of these shapes.
All lengths are in centimetres.

a

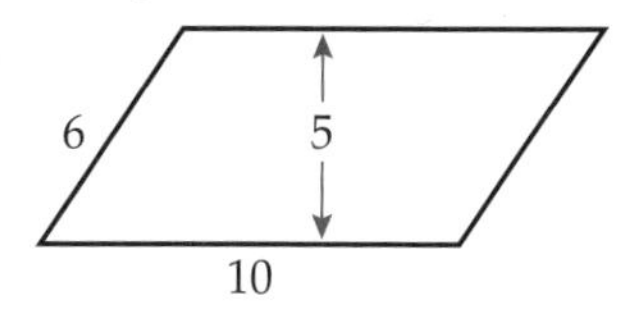

b

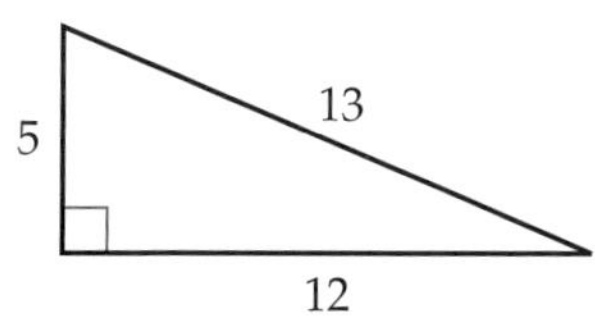

D

3 This parallelogram has an area of 45 cm^2.
Work out the height, h cm, of the parallelogram.

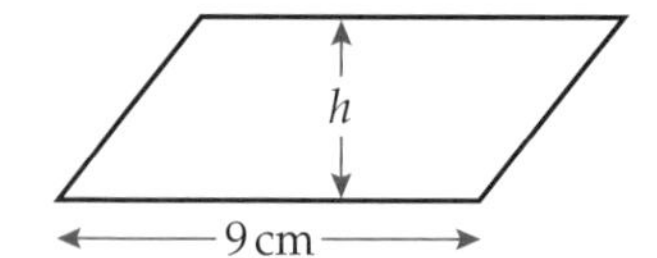

4 This triangle has an area of 15 cm^2.
Work out the height, h cm, of the triangle.

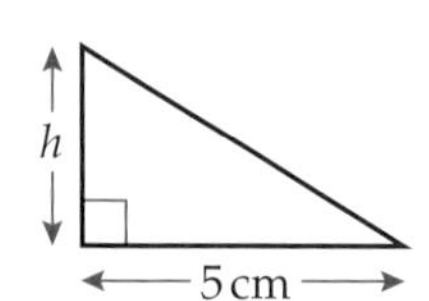

AO2

D

5 Calculate the areas of these trapezia.
All lengths are in centimetres.

a

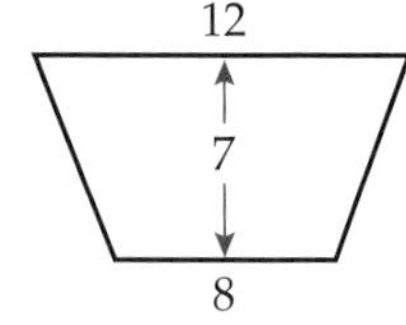

b

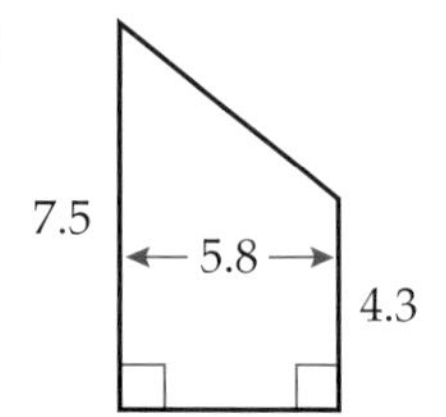

D

6 Caroline has a rectangular dining room that measures 3.2 m by 4 m.
She buys laminate flooring that costs £11.95 per square metre.
She can only buy the flooring in whole numbers of square metres.
Delivery, underlay and fitting costs £57.50.
How much does it cost Caroline for the flooring, delivery, underlay and fitting?

7 Which shape has the larger area, and by how much?

A

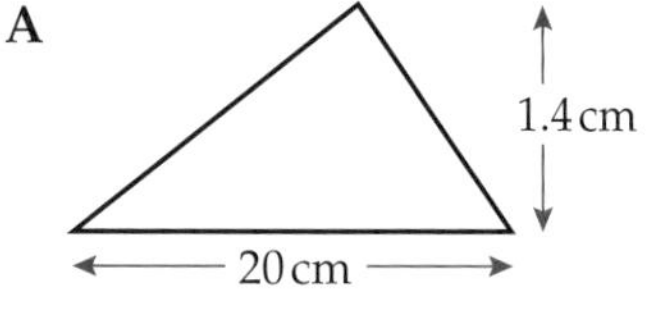

B

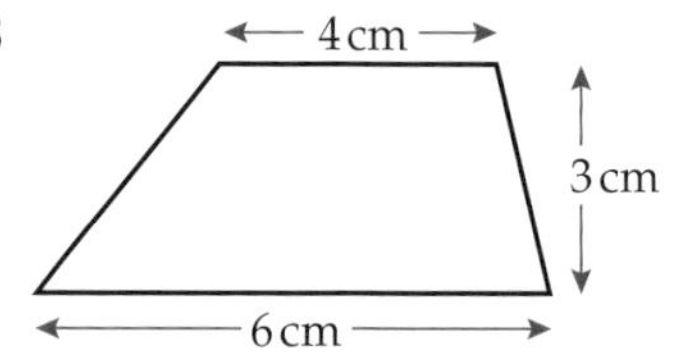

AO2

26.2 Perimeter and area of compound shapes

D

1 **a** Calculate the perimeter of this compound shape.

b Calculate the area of this compound shape by first dividing the shape into two rectangles, A and B.

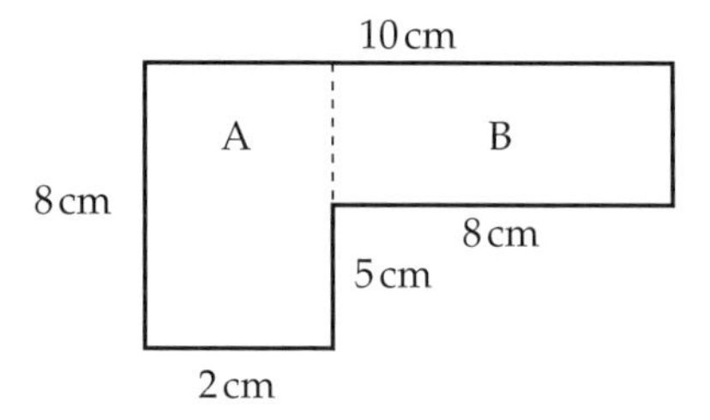

2 Calculate the perimeter and area of each of these compound shapes.
All lengths are in centimetres.

a

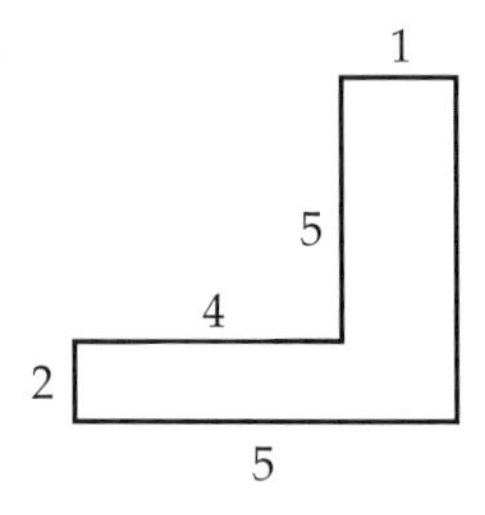

b

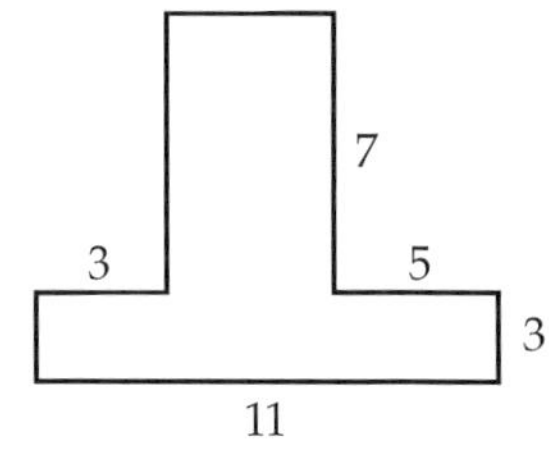

3 Calculate the area of each of these compound shapes.
All lengths are in centimetres.

a

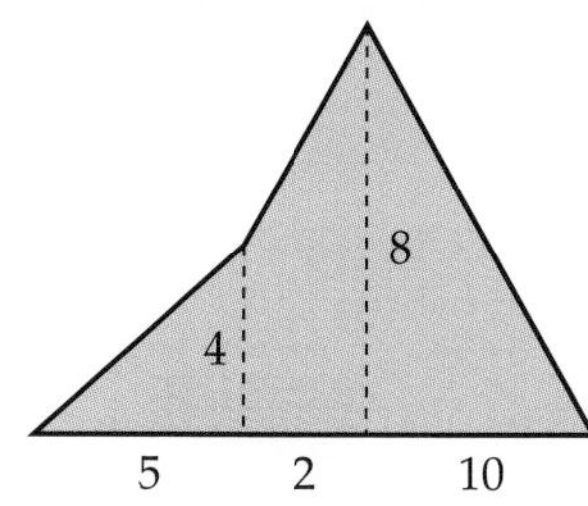

b

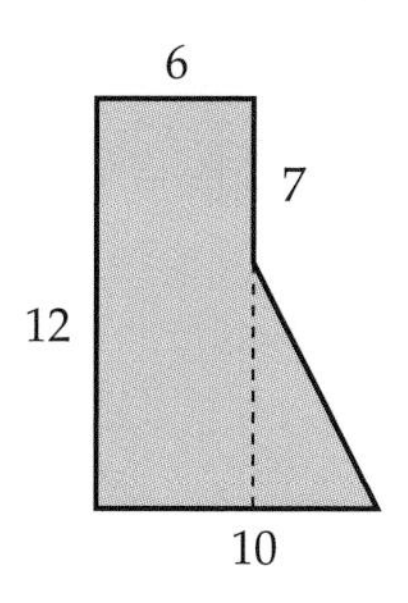

D

26.3 Volume and surface area of prisms

1 Calculate the volume of this prism.

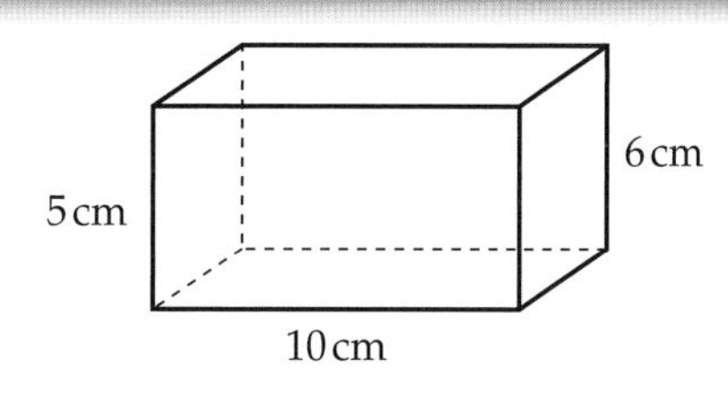

E

2 Calculate the volumes of these prisms. All measurements are in centimetres.

a

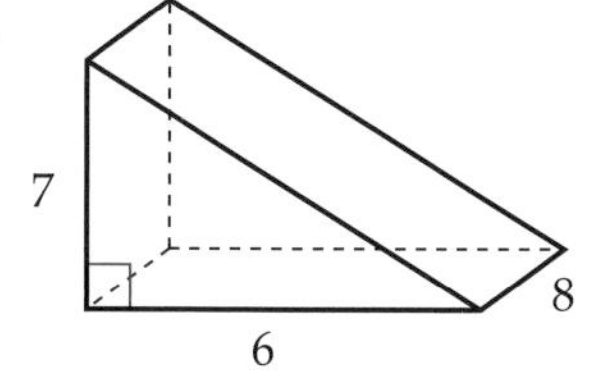

b

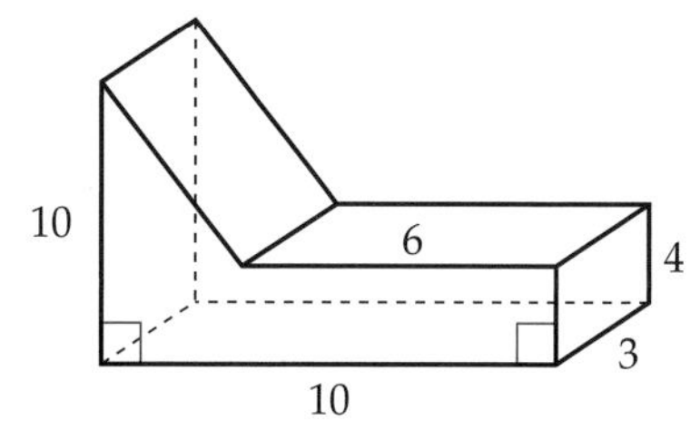

D C

3 Harry has a glass conservatory on the side of his house.
Harry's conservatory is 4.5 m long.
What is the volume of the conservatory?

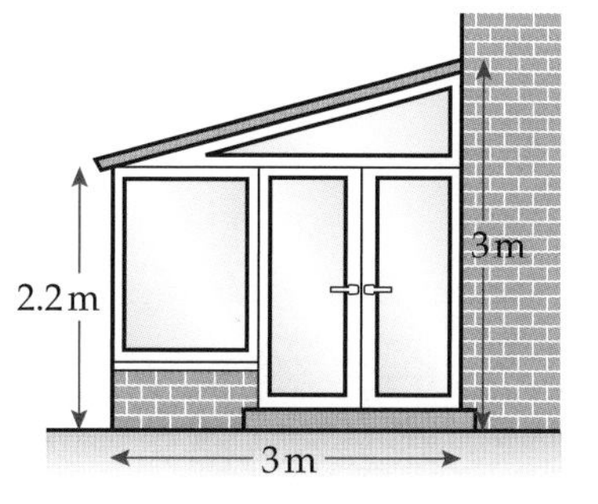

C AO2

4 Trina is the manager of a swimming pool.
The dimensions of the swimming pool are shown on the diagram.
To clean the pool Trina adds 100 g of chlorine for every 50 m^3 of water.
How many kilograms of chlorine must Trina add to the pool?

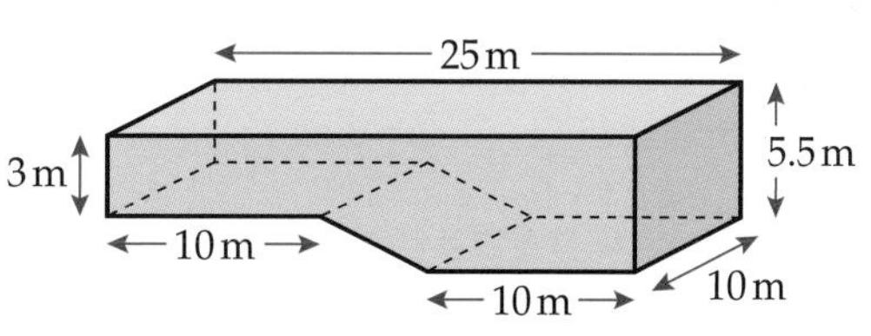

C AO3

5 Calculate the surface area of each of these prisms.

a

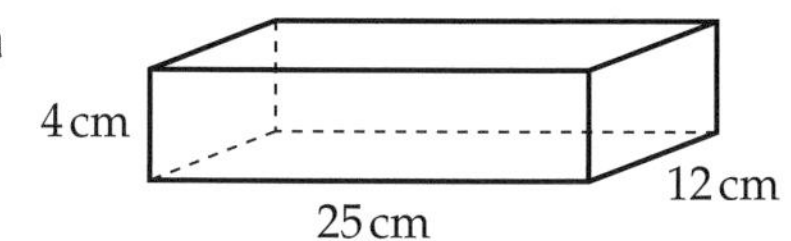

b

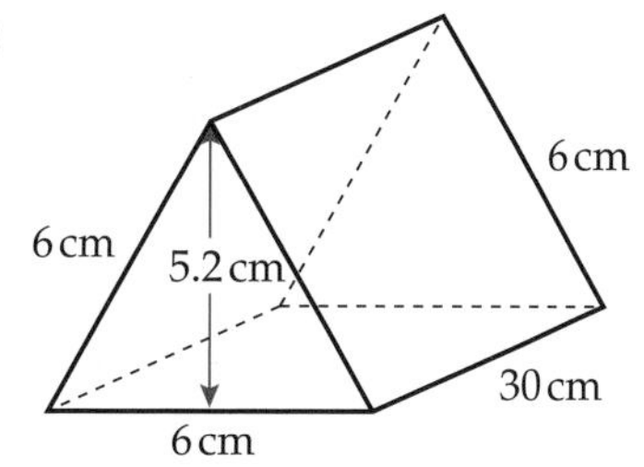

D C

Links to:
Middle Student Book
Ch27, pp.416–421

Key Points

Isometric drawing (E)

Draw along the printed lines of the paper.

Vertical lines on the paper represent the vertical lines of the object.

The lines at an angle on the paper represent the horizontal lines on the object.

Plans and elevations (D)

The plan is the view from above the object.

The front elevation is the view from the front of the object.

The side elevation is the view from the side of the object.

Plane of symmetry (D)

When a 3-D object is cut along a plane of symmetry, it will be cut into two identical halves.

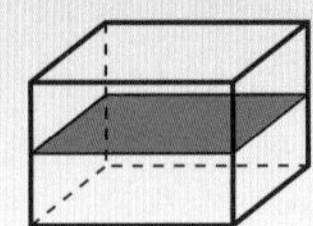
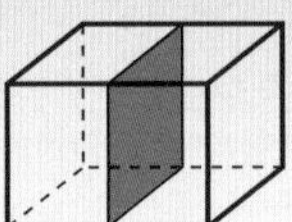
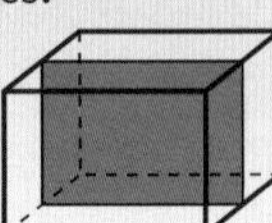

27.1 Drawing 3-D objects

E

1 On isometric paper draw 3-D diagrams of the objects with these cross-sections.
Assume that the objects are 4 cm wide.

a
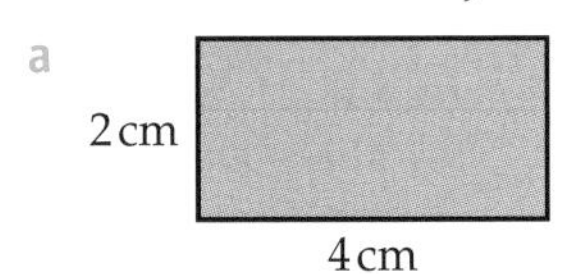

b
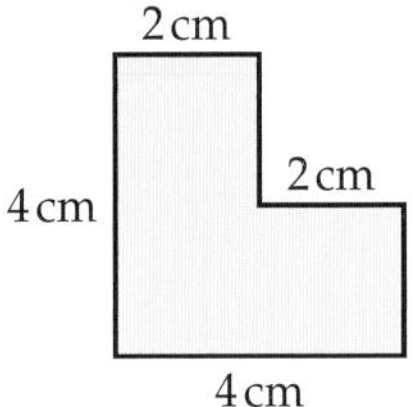

c
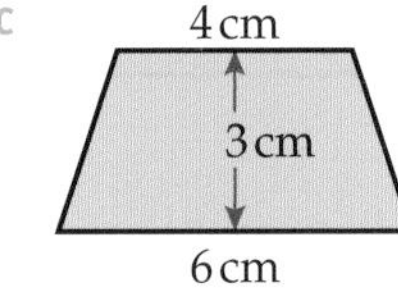

D

2 For each block of cubes, draw

a a plan view

b a front elevation

c a side elevation from the right-hand side.

A
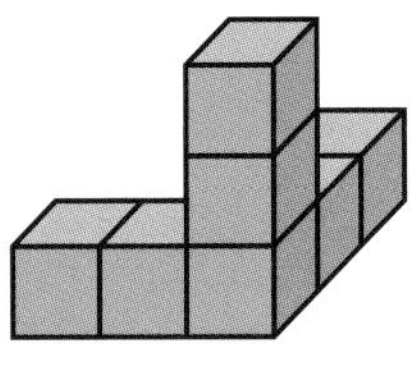
7 cubes

B
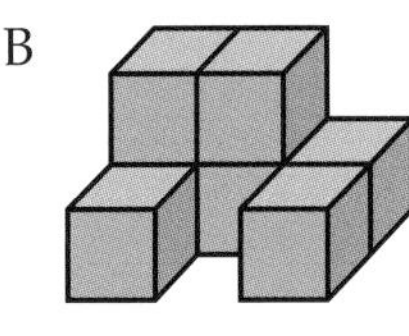
7 cubes

C
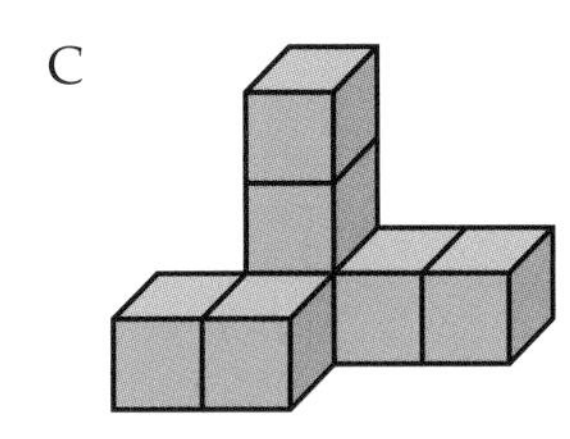
7 cubes

3 Work out the volume and surface area of each block of cubes in Q2.
All the cubes have side 1 cm.

4 Copy these 3-D objects onto squared paper.
For each object show all its planes of symmetry.
Draw separate diagrams for each plane of symmetry.

a
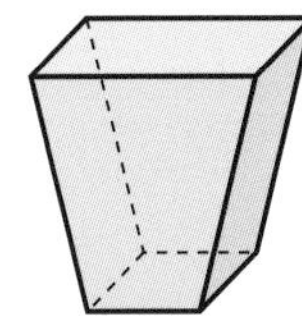

b
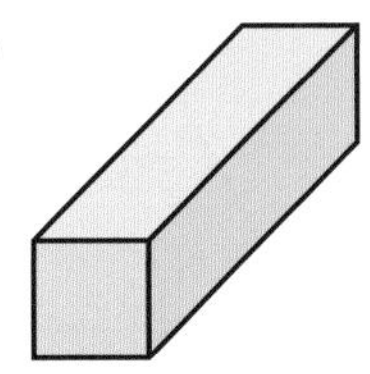

Reflection, translation and rotation

Links to: Middle Student Book Ch28, pp.422–442

Key Points

Reflection

E D C

In a reflection, the object and the image are the same perpendicular distance from the mirror line, on opposite sides.

To describe a reflection on a grid you need to give the equation of the mirror line.

Translation

D C

To describe a translation you have to give the distance and direction of movement.

You can describe a translation using a column vector.

$\begin{pmatrix}3\\2\end{pmatrix}$ means move 3 in the x-direction and then move 2 in the y-direction.

Rotation

D C

To describe a rotation fully, you need to give
- the centre of the rotation
- the size of turn
- the direction of turn.

Combined transformations

C

Reflection, translation and rotation are transformations. They transform an object to an image. Combined transformations can often be described in terms of a single transformation.

28.1 Reflection on a coordinate grid

E

1 Copy this coordinate grid.

a Draw the reflection of triangle A in the y-axis. Label the reflected shape B.

b Draw the reflection of triangle A in the x-axis. Label the reflected shape C.

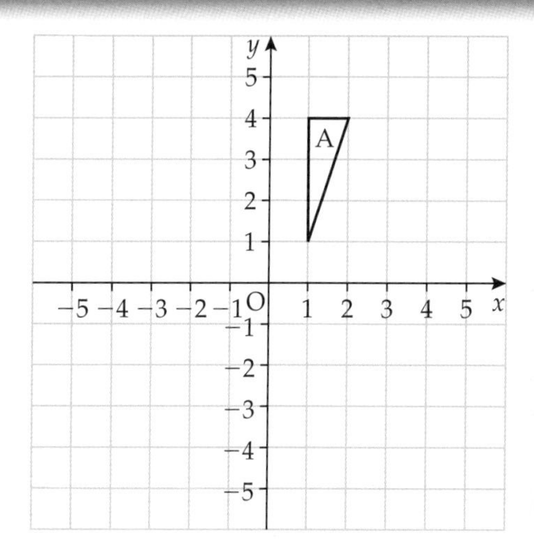

D

2 Copy this coordinate grid.

a Draw the reflection of shape T in the line $x = 3$. Label the reflected shape U.

b Draw the reflection of shape T in the line $x = 1$. Label the reflected shape V.

c Draw the reflection of shape Q in the line $y = -1$. Label the reflected shape R.

d Shape S is a reflection of shape Q in a mirror line. Describe this transformation.

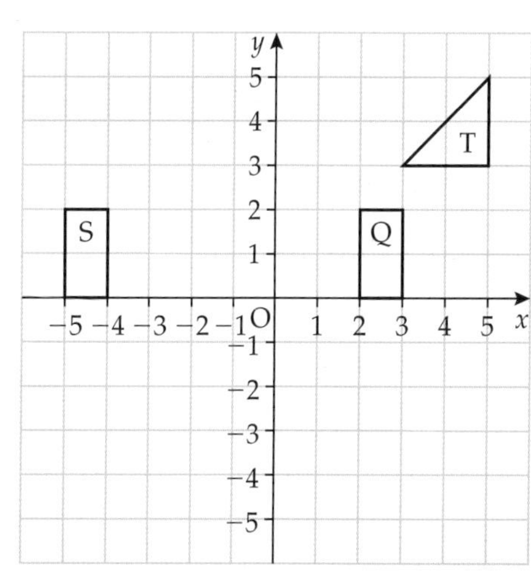

AO2

C

3 Copy this coordinate grid.

a Draw the line $y = x$ on your grid using a dotted line.

b Draw the reflection of shape A in the line $y = x$.
Label the reflected shape D.

c Draw the reflection of shape B in the line $y = x$.
Label the reflected shape E.

d Draw the reflection of shape C in the line $y = x$.
Label the reflected shape F.

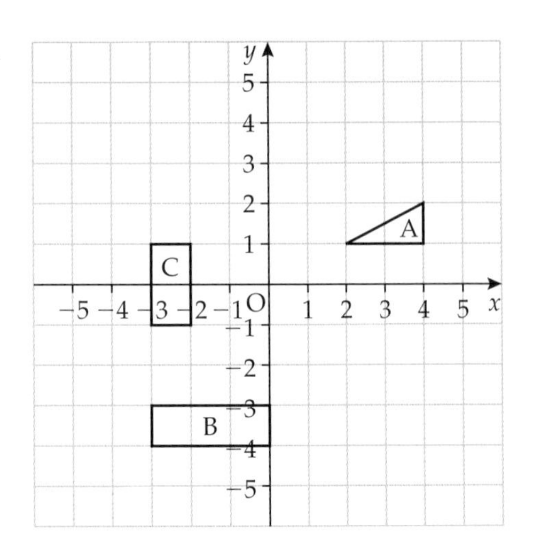

C

4 Rodrigo starts with this shape.

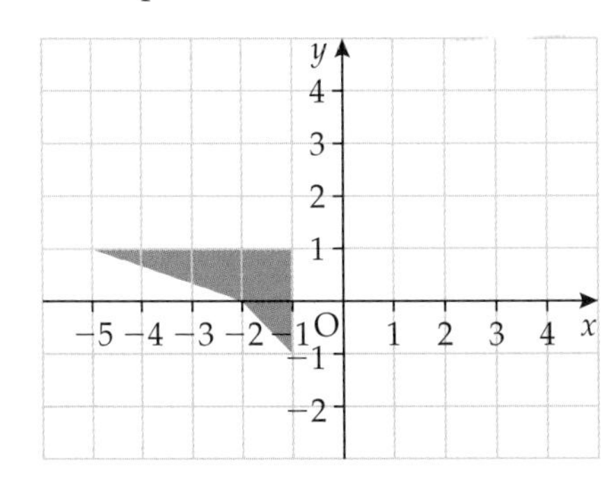

He transforms the shape to make this pattern.

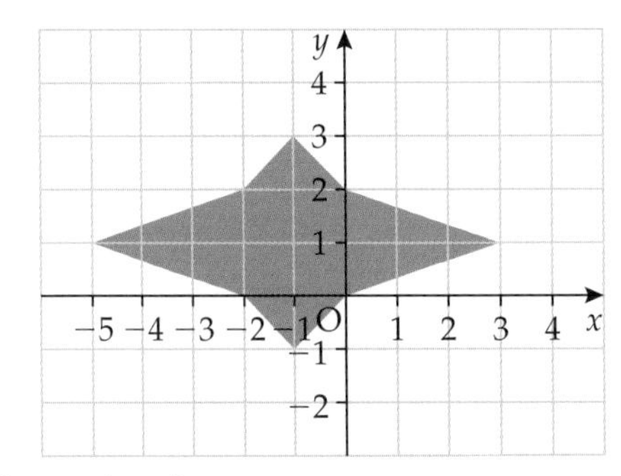

AO3

Describe the transformations he uses.

28.2 Translation

D

1 Copy the following shapes on squared paper.
Draw the image of the shape after the given translation.

a

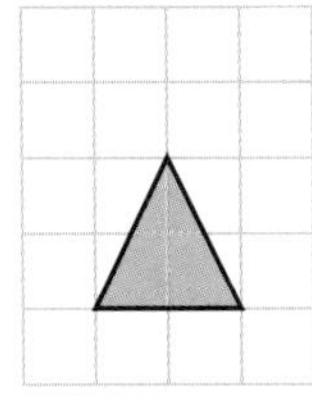

2 squares right
1 square up

b

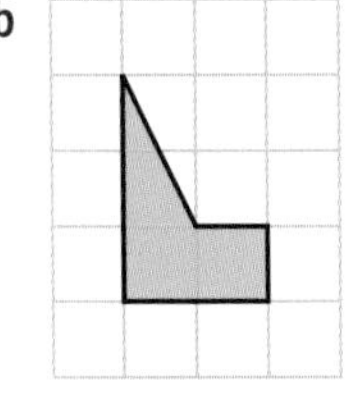

1 square left
3 squares up

c

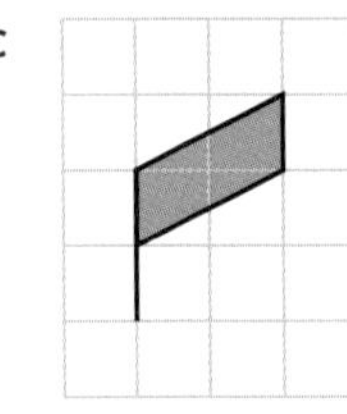

2 squares left
3 squares down

2 Describe the translation that takes

a shape A to shape B

b shape A to shape D

c shape A to shape C

d shape C to shape A.

e What do you notice about your answers to parts **c** and **d**?

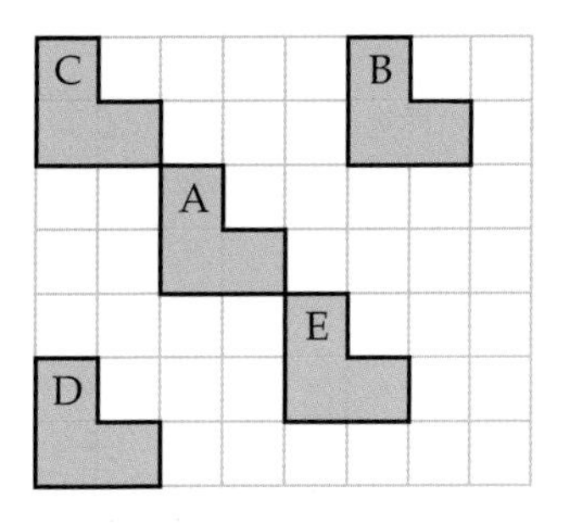

D

AO2

3 Look at the grid used in Q2.

Write down the column vector for each of the following translations.

a shape A to shape B

b shape B to shape D

c shape E to shape D

d shape C to shape E

e shape E to shape C

f What do you notice about your answers to parts **d** and **e**?

The translation that takes shape D to shape A is $\begin{pmatrix} 2 \\ 3 \end{pmatrix}$.

g Write down the column vector that takes shape A to shape D.

C

AO2

4 On a grid, shape P is translated to shape Q by the vector $\begin{pmatrix} 3 \\ -4 \end{pmatrix}$ and shape Q is translated to shape R by the vector $\begin{pmatrix} 1 \\ 1 \end{pmatrix}$.

Write down the column vector that translates shape P directly to shape R.

C

AO3

28.3 Rotation

1 For each part of this question, copy this shape and the centre of rotation on squared paper.

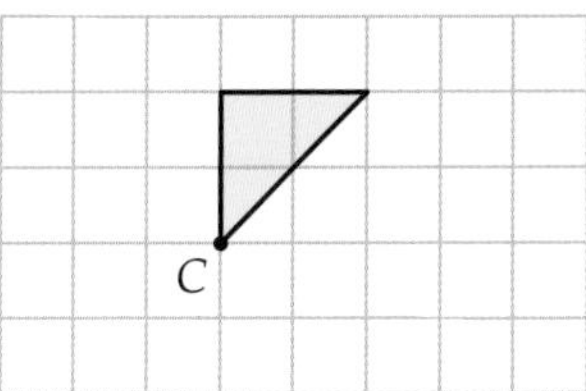

Draw the image of the shape after a rotation of

a half a turn about centre C

b 90° anticlockwise about centre C.

D

2 Copy this shape on squared paper.
Follow the instructions to make a pattern.

- Rotate the shape a quarter turn clockwise about the centre C.
- Draw the image.
- Rotate the image a quarter turn clockwise about the centre C.
- Draw the image.
- Repeat until the pattern is complete.

What is the order of rotational symmetry of the pattern?

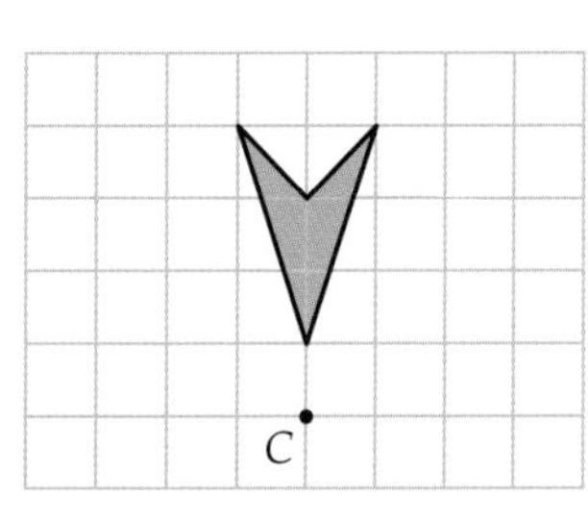

D

AO2

C

3 Copy this pattern on a coordinate grid.
Rotate the pattern about centre (0, 0) to make a pattern with rotational symmetry of order 4.

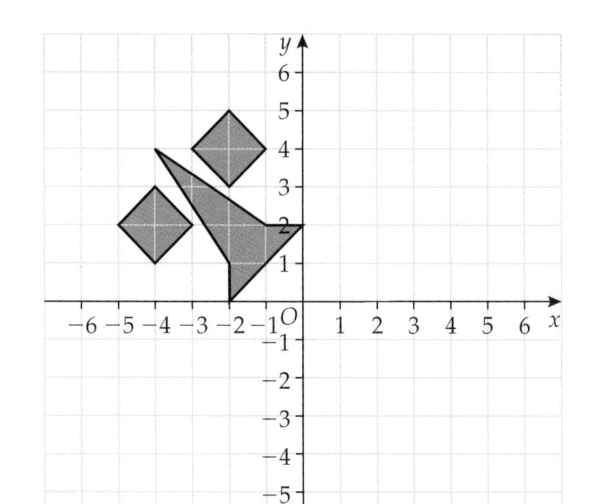

AO2

Q4 and Q5 use this coordinate grid.

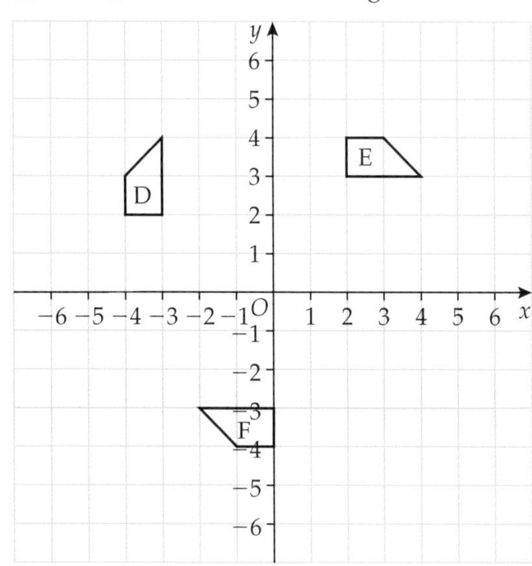

D

4 Shape D rotates onto shape E.

a What size turn is the rotation?

b Find the centre of rotation and write down the coordinates.

C

5 Describe fully the transformation that maps shape E onto shape F.

AO2

C

6 A children's fairground wheel has eight seats.
The seats are at the end of arms which are spaced equally around a centre hub.
Each seat and arm look like this.

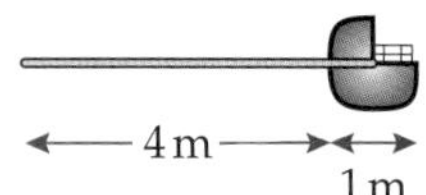

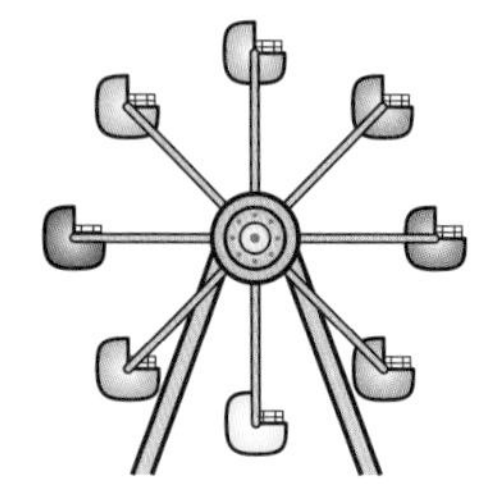

Make a simplified scale drawing of the wheel on a coordinate grid.
Use a scale of 1 grid square to 1 metre.

AO3

Copy this diagram only once. Use the same coordinate grid to answer all of the questions.

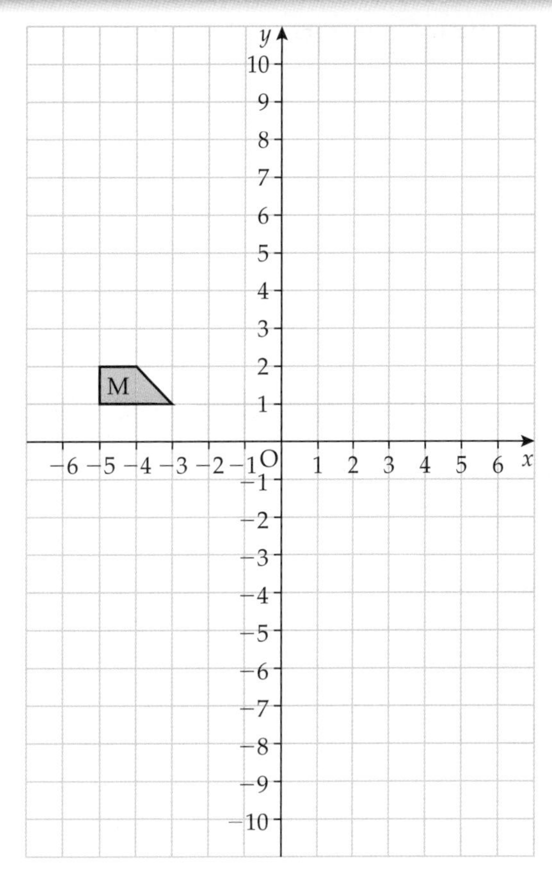

C

1 a Translate shape M by column vector $\begin{pmatrix} 2 \\ 4 \end{pmatrix}$. Label the image N.

b Translate shape N by column vector $\begin{pmatrix} -2 \\ 0 \end{pmatrix}$. Label the image L.

c Write down the column vector for the transformation that takes shape M directly to shape L.

2 a Rotate shape M 90° anticlockwise about (0, 0). Label the image P.

b Rotate shape P 90° anticlockwise about (1, 1). Label the image Q.

c Describe the single transformation that takes shape M to shape Q.

3 a Reflect shape M in the y-axis. Label the image R.

b Reflect shape R in the line $x = 3$. Label the image S.

c Reflect shape S in the line $x = 1$. Label the image T.

d Describe the single transformation that takes shape M to shape T.

4 a Rotate shape M 90° clockwise about (−3, 1). Label the image U.

b Translate shape U by column vector $\begin{pmatrix} 5 \\ 1 \end{pmatrix}$. Label the image V.

c Describe the single transformation that takes shape M to shape V.

C

5 Starting with shape M, describe four of your own transformations that do the following:

- 1st transformation rotates shape M to shape W
- 2nd transformation translates shape W to shape X
- 3rd transformation rotates shape X to shape Y
- 4th transformation translates shape Y to shape M.

Your shapes W, X, and Y must not overlap with any of the other shapes drawn on your coordinate grid.

A02

Links to:
Middle Student Book
Ch29, pp.443–456

Key Points

Circle

D C

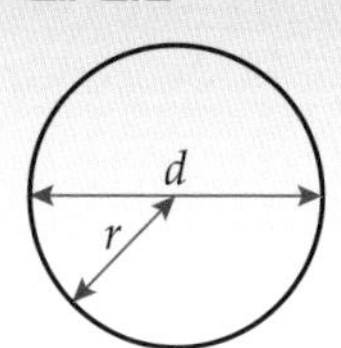

Circumference = $2\pi r$ or πd

Area = πr^2

Cylinder

C

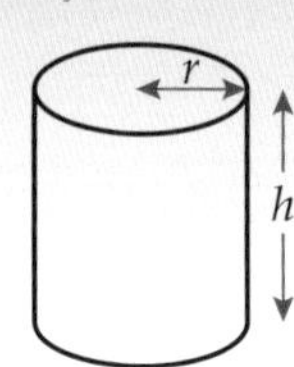

Volume = $\pi r^2 h$

Curved surface area = $2\pi rh$

Total surface area = $2\pi r(h + r)$

29.1 Circumference of a circle

For these questions, use the π button on your calculator unless you are asked to leave your answer in terms of π.

D

1 Calculate the circumference of each circle.
Give your answer correct to 1 d.p.

a radius = 8.4 cm **b** diameter = 4.8 cm

2 Calculate the circumference of a circle with a radius of 1.25 m.
Leave your answer in terms of π.

D A02

3 A car tyre has a diameter of 66 cm.
How many complete revolutions does the wheel make on a journey of 50 km?

D

4 **a** Calculate the diameter of a circle with a circumference of 100 cm.
Give your answer correct to 2 s.f.

b Calculate the radius of a circle with a circumference of 100π cm.

C

5 Calculate the perimeter of this paving slab.
Give your answer to the nearest centimetre.

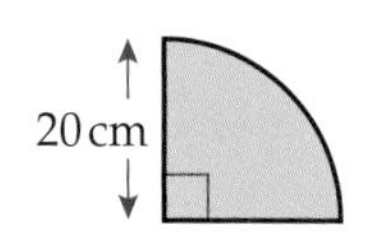

C A02

6 A circular pattern made from paving slabs is shown.
The thick lines are where cement has to be placed, to lock the slabs together.
The cement is called grout.
Work out the total length of grout needed for this pattern of slabs.
Give your answer to the nearest centimetre.

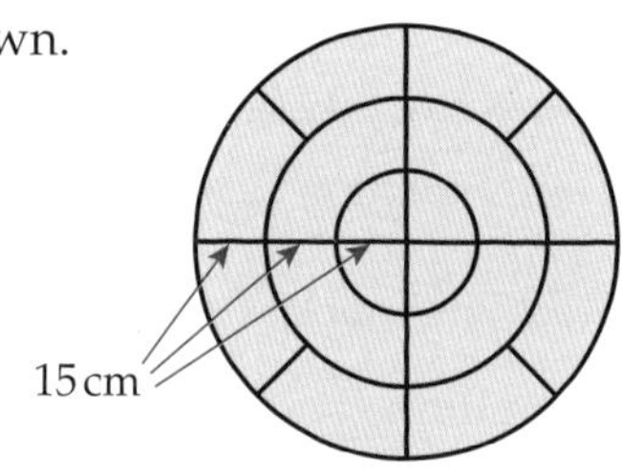

29.2 Area of a circle

D

1 Calculate the area of a circle with a diameter of 11 cm.
Leave your answer in terms of π.

2 Calculate the radius of a circle with an area of 16π cm^2.

3 Calculate the area of each shape. Give your answer to 1 d.p.

a

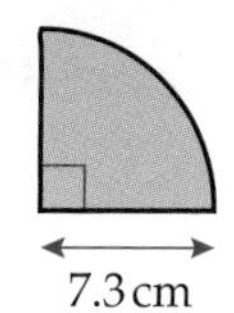

b

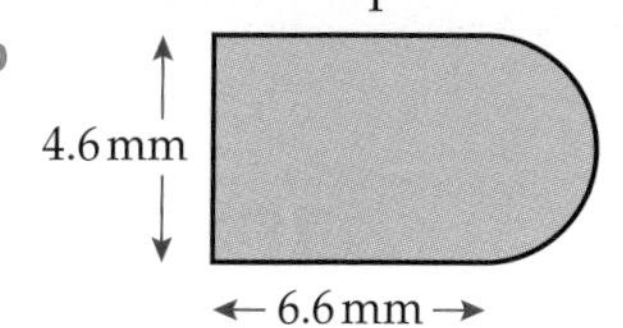

4 Calculate the shaded area in each of the following shapes.

a

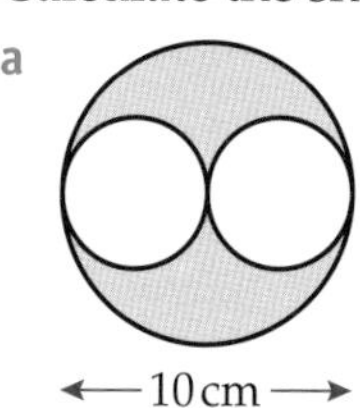

b

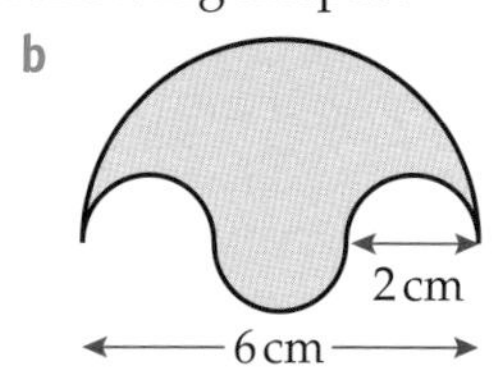

5 A semicircular piece of steel has a diameter of 30 cm.
Enamel costs 17p per square centimetre.
Work out the cost of painting one side of the piece of steel with enamel.

29.3 Cylinders

1 Calculate the volume of this cylinder.
Leave your answer in terms of π.

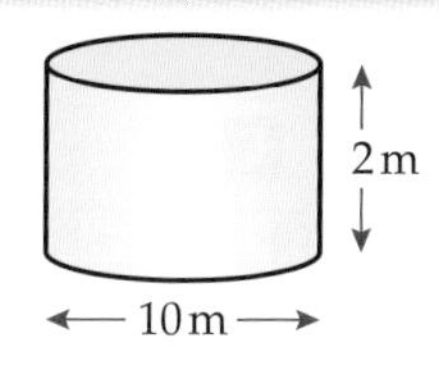

2 Calculate the volume of a cylinder with a radius of 10 cm and a height of 6 cm.
Give your answer correct to 2 d.p.

3 An energy drink is sold in cans with a radius of 2.6 cm and a height of 13.5 cm.
As many cans of the drink as possible are packed into this box.

a What is the volume of one can?
Give your answer in terms of π.

b What volume of the box is empty?
Give your answer to the nearest whole number.

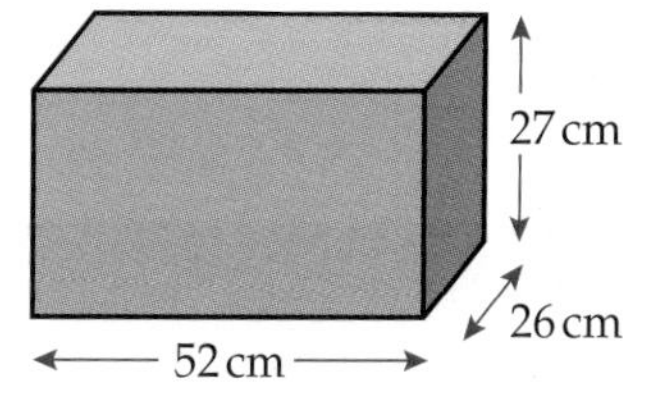

4 A boiler contains hot water.
A pipe connecting the boiler to a tap is 14.5 m long and has a diameter of 1.5 cm.
When the tap is switched on, the cold water from the pipe pours into the sink before the hot water arrives.
How many litres of cold water pour into the sink before the hot water arrives?

1 litre = 1000 cm^3

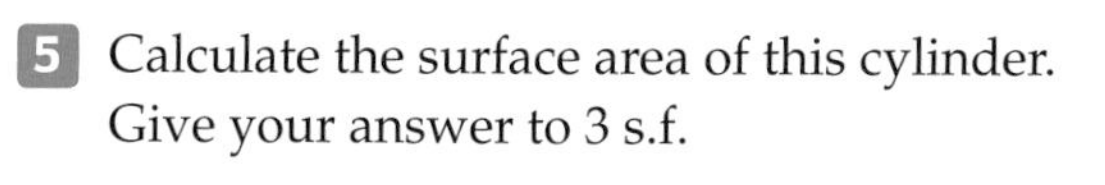

5 Calculate the surface area of this cylinder.
Give your answer to 3 s.f.

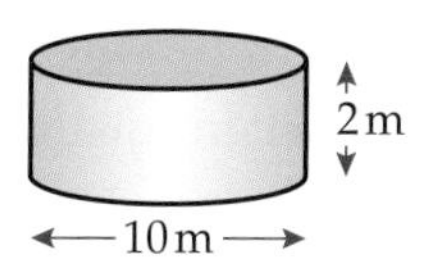

6 Which of these two cylinders has the larger surface area, and by how much?
Give your answers to 4 s.f.

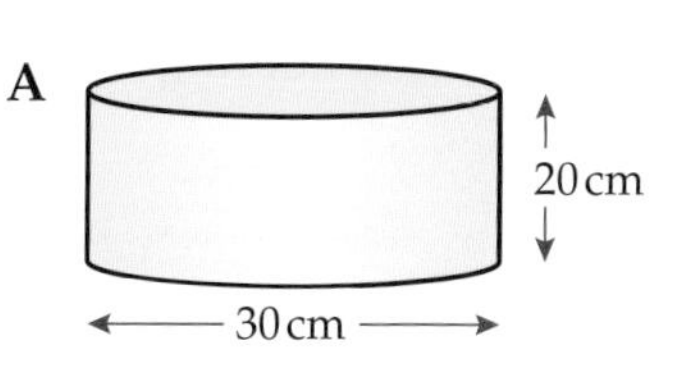

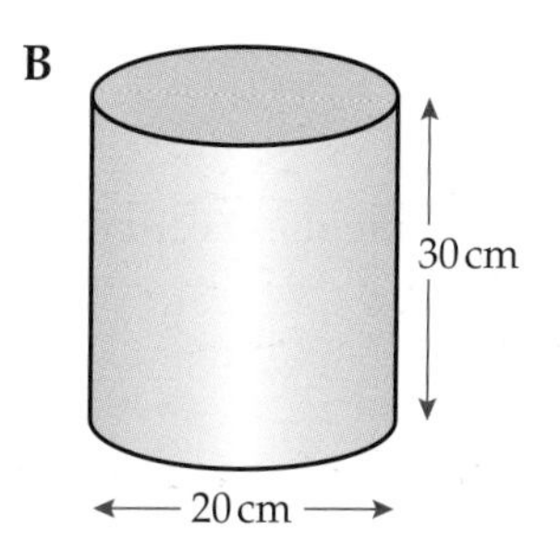

30 Measurement 2

Links to:
Middle Student Book
Ch30, pp.460–471

Key Points

Converting areas and volumes D C

You can convert units of area using:

$1\text{ cm}^2 = 100\text{ mm}^2$

$1\text{ m}^2 = 10\,000\text{ cm}^2$

$1\text{ km}^2 = 1\,000\,000\text{ m}^2$

You can convert units of volume using:

$1\text{ cm}^3 = 1000\text{ mm}^3$

$1\text{ m}^3 = 1\,000\,000\text{ cm}^3$

$1\text{ litre} = 1000\text{ cm}^3$

Speed D

Speed is a measurement of how fast something is travelling.

$\text{speed} = \dfrac{\text{distance}}{\text{time}}$

$\text{distance} = \text{speed} \times \text{time}$

$\text{time} = \dfrac{\text{distance}}{\text{speed}}$

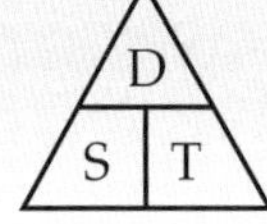

Density C

Density is a measurement of the amount of a substance contained in a certain volume.

$\text{density} = \dfrac{\text{mass}}{\text{volume}}$

$\text{mass} = \text{density} \times \text{volume}$

$\text{volume} = \dfrac{\text{mass}}{\text{density}}$

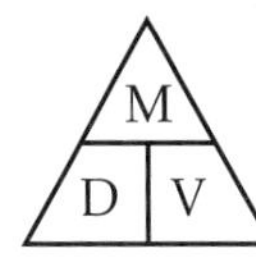

Dimension theory B

Length is a one-dimensional quantity.

In a formula for length, each term must contain one letter representing length.

Area is a two-dimensional quantity.

In a formula for area, each term must contain either two lengths multiplied together or one letter representing area.

Volume is a three-dimensional quantity.

In a formula for volume, each term must contain either three lengths multiplied together, a length multiplied by an area or one letter representing volume.

30.1 Converting areas and volumes

D

1 Copy and complete these conversions.

a $1\text{ m}^2 = \square\text{ cm}^2$ b $1\text{ cm}^2 = \square\text{ mm}^2$ c $1\text{ km}^2 = \square\text{ m}^2$

d $2.5\text{ m}^2 = \square\text{ cm}^2$ e $12\text{ cm}^2 = \square\text{ mm}^2$ f $0.75\text{ km}^2 = \square\text{ m}^2$

g $750\text{ cm}^2 = \square\text{ m}^2$ h $140\text{ mm}^2 = \square\text{ cm}^2$ i $1250\text{ m}^2 = \square\text{ km}^2$

D A02

2 Phillip's kitchen is a rectangle 220 cm wide by 630 cm long.

a Work out the floor area of Phillip's kitchen in square metres.
Give your answer to the nearest square metre.

b Cushion flooring costs £19.99 per square metre.
How much does it cost Phillip to buy the cushion flooring he needs for his kitchen floor?

C

3 Copy and complete these conversions.

a $1\text{ m}^3 = \square\text{ cm}^3$ b $1\text{ cm}^3 = \square\text{ mm}^3$ c $1\text{ km}^3 = \square\text{ m}^3$

d $1\text{ litre} = \square\text{ cm}^3$ e $0.3\text{ m}^3 = \square\text{ cm}^3$ f $0.04\text{ km}^3 = \square\text{ m}^3$

g $7500\text{ cm}^3 = \square\text{ litres}$ h $15\,000\text{ mm}^3 = \square\text{ cm}^3$ i $7500\text{ cm}^3 = \square\text{ m}^3$

4 A bucket containing 4 litres of water has a hole in the bottom.
Water leaks out of the hole at a rate of $5\,\text{mm}^3$ per second.

a Convert 4 litres into mm^3.

b Assuming the leak continues at the same rate, how long will it take for the bucket to empty? Give your answer to the nearest day.

C

AO3

30.2 Speed

D

1 Work out the average speed for each of these journeys.

a A tuna that swims 90 m in 6 seconds.

b A turtle that swims 90 m in $1\frac{1}{2}$ minutes.

c Captain Webb was the first person to swim the 35 km across the English Channel in 1875. He took 21.75 hours.

2 Work out the distance travelled for each of these journeys.

a A remote-controlled plane that travels at 27 mph for half an hour.

b A leaf that falls at 1.2 m/s for 7 seconds.

Remember that m/s means metres per second.

3 Work out the time taken for each of these journeys.

a A car that travels 280 miles at an average speed of 56 mph.

b A paper plane that travels 30 m at an average speed of 1.6 m/s.

D

AO2

4 The speed limit on many streets is 20 mph.

a Convert 20 miles to kilometres.

b Write down this speed limit in km/h.

FUNCTIONAL

D

5 Brian is driving through France on his holiday.
The speed limit is 120 km/h.
Brian is travelling at 80 mph.
Is Brian breaking the speed limit?
Show working to support your answer.

Start by converting 80 miles into kilometres.

6 Pierre is driving through England on his holidays.
The speed limit is 60 mph.
Pierre is travelling at 90 km/h.
Is Pierre breaking the speed limit?
Show working to support your answer.

AO3

30.3 Density

C

1 In a physics experiment, Leanne records the volume and mass of three different types of glass.
The table shows her results.
Work out the density of each type of glass.
Give your answer correct to 1 d.p.

Type of glass	Volume (cm^3)	Mass (g)
cheap	35	74
lead crystal	17	53
optical	3.5	25

C

2 Bullet-proof glass has a density of 2.23 g/cm^3.
The diagram shows the dimensions of a bullet-proof window.

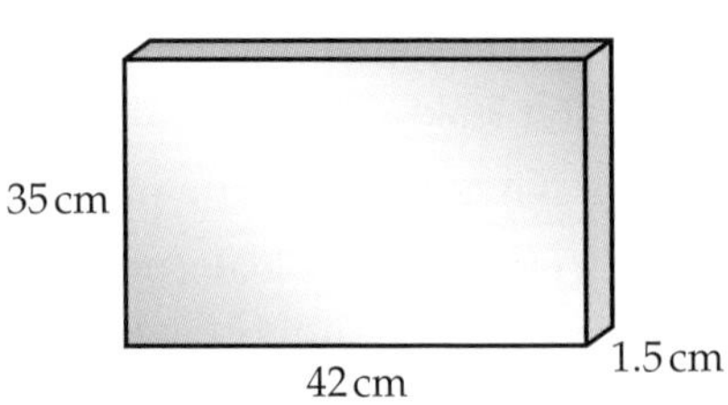

a Calculate the volume of the window.

b Work out the mass of the window.
Give your answer to the nearest gram.

AO2

C

3 This old oak beam has a cross-section area of 0.08 m^2.
It weighs 500 kg.
Work out the density of the oak beam.

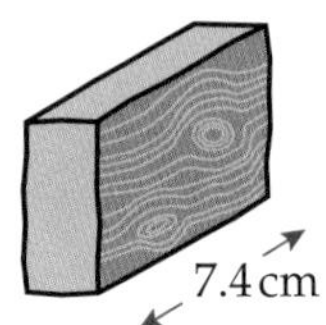

4 A large gold cross is made from 35.2 cm^3 of 14 carat gold.
The density of 14 carat gold is 13.2 g/cm^3.
The scrap metal value of 14 carat gold is £11.66 per gram.
Work out the scrap metal value of the gold cross.

AO3

30.4 Dimension theory

B

1 E, F and G represent areas, and r and s represent lengths.
Write down whether each of these expressions represents a length (L), an area (A), a volume (V) or none of these (N).

a $E + F + G$ b $0.3rs$ c πr^2 d $\frac{4}{3}\pi Es$

e $E + 3rs$ f $EF - G$ g $E + \pi r^3$ h $rE + sF$

2 In these expressions a, b and c all represent lengths.
Write down whether each of these expressions represents a length (L), an area (A), a volume (V) or none of these (N).

a $a + c$ b bc c abc d a^2c

e $\pi a^2 - bc$ f $2a + b^2$ g $ab + c$ h $2a + \pi b$

i $3a(a + c)$ j $\pi(a + b)$ k $3a(a^2b - abc)$ l $a^2\pi(\pi a + 3c)$

B

3 Adam writes down the following formula for the volume of a cone of radius r and height h.

$$V = \frac{1}{3}\pi r^2 h^2$$

Use dimensions to show why Adam's formula must be wrong.

4 In these expressions w is measured in cm, x is measured in cm^2 and y is measured in cm^3.

Write down the units you would use for each expression.

a wx b $\frac{y}{w}$ c $\frac{x}{w}$ d $\frac{y}{w^2}$ e $2\pi wx - 5wy$

AO2

B

5 A feng shui advisor suggests putting fresh oranges in a round glass bowl at the centre of a dining table.

The advisor uses a formula of $\frac{r^2}{10}$ for the number of oranges to put in different size bowls, where r is the radius of the bowl.

Do you think the formula will provide a useful guide for all the different sizes of bowls?

Give a reason for your answer.

AO3

31 Enlargement and similarity

Links to:
Middle Student Book
Ch31, pp.474–485

Key Points

Enlargement

E D C

An enlargement changes the size of an object but not its shape.

The number of times each length of a shape is enlarged is called the scale factor.

In an enlargement, all the angles stay the same but all the lengths change in the same proportion. The image is similar to the object.

To describe an enlargement fully you must give the scale factor and the centre of enlargement.

A scale factor greater than 1 gives an image that is larger than the object.

A scale factor between 0 and 1 gives an image that is smaller than the object.

Similarity

B

Two objects are similar when they are exactly the same shape but not the same size.

An enlargement always produces two shapes that are similar.

For two shapes to be similar
- the corresponding angles must be equal
- the ratios of corresponding sides must be the same.

31.1 Enlargement

E

1 a Copy this shape on squared paper.

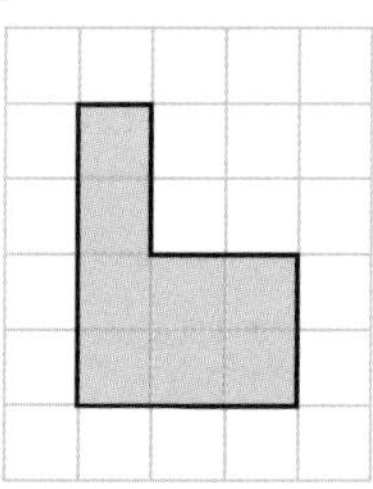

b Enlarge the shape by a scale factor of 3.

c What is the perimeter of the object?

d What is the perimeter of the image?

e How many times larger is the perimeter of the image than the perimeter of the object?

D

2 Copy each of these shapes on squared paper. Draw the enlargement of each shape using the scale factor given and point O as the centre of enlargement.

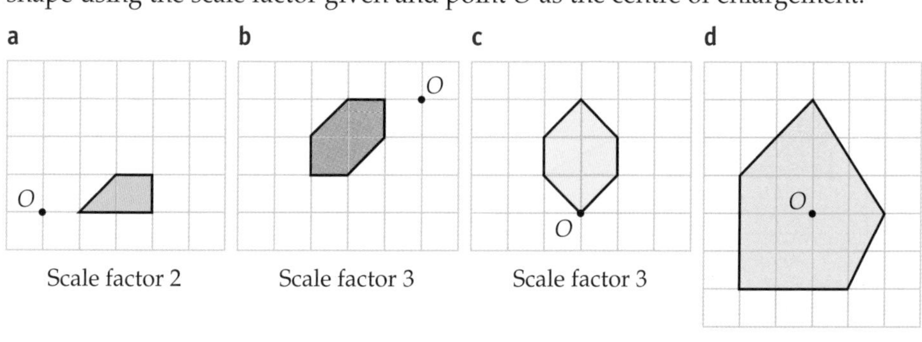

a Scale factor 2 b Scale factor 3 c Scale factor 3 d Scale factor 4

D

3 The diagram shows two shapes A and B on a grid. Shape B is an enlargement of shape A.

a What are the coordinates of the centre of enlargement?

b What is the scale factor of the enlargement?

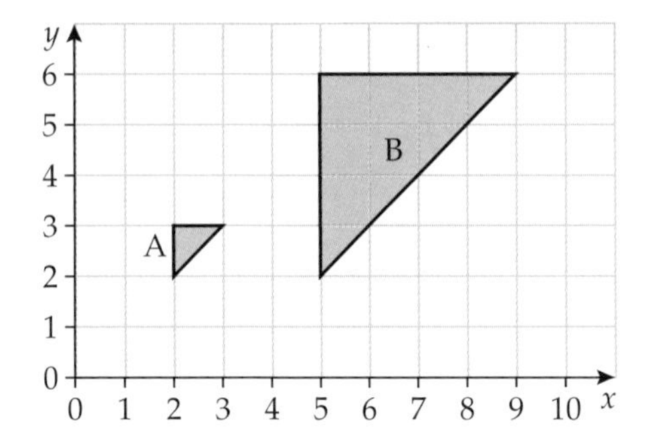

4 On squared paper draw a coordinate grid with x- and y-axes from 0 to 6. Plot a triangle with vertices at (2, 1), (2, 4) and (4, 1). Label the triangle T.

a Draw the image of T after an enlargement with scale factor 2, using (3, 2) as the centre of enlargement. Label the image S.

b What are the coordinates of the vertices of S?

5 A quadrilateral has a perimeter of 9 cm.
It is enlarged by a scale factor of 20.
What is the perimeter of the enlarged shape?

31.2 Enlargement with fractional scale factors

C

1 Copy this diagram on squared paper.

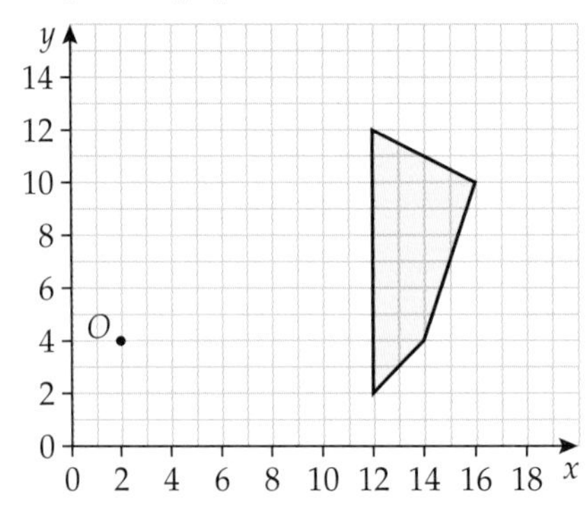

a Enlarge the shape using a scale factor of $\frac{1}{2}$ and centre of enlargement O.

b What enlargement would take the image back to the object?

2 Copy this diagram on squared paper.

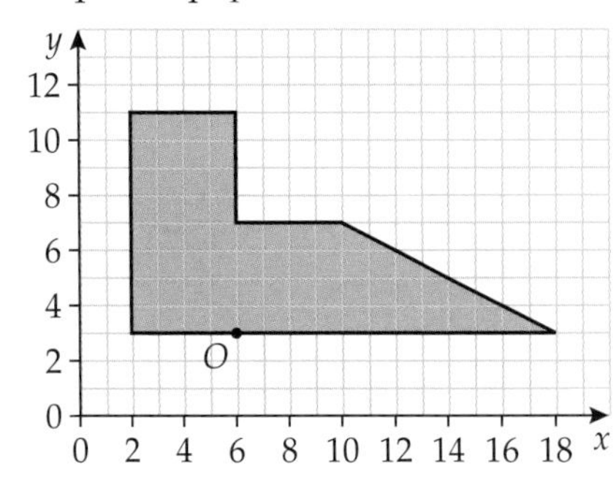

a Enlarge the shape using a scale factor of $\frac{3}{4}$ and centre of enlargement O.

b What enlargement would take the image back to the object?

3 Shape A is an enlargement of shape B.
Copy the diagram on squared paper.

a What is the scale factor of the enlargement?

b Use construction lines to find the centre of enlargement.
Write down the coordinates of the centre of enlargement.

c What enlargement would take shape B back to shape A?

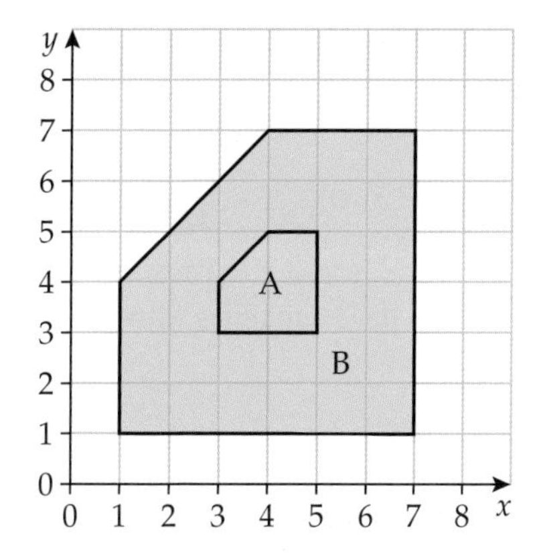

C

31.3 Similarity

B

1 Triangle T is an enlargement of triangle S.

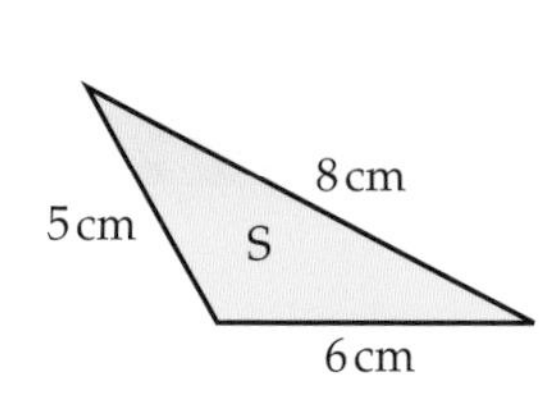

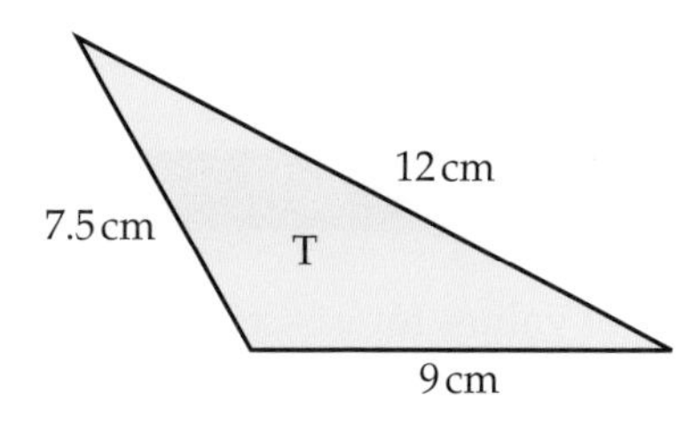

a What is the ratio of the longest sides of the triangles?
Write your answer in its simplest form.

b What is the ratio of the perimeters of the triangles?
Write your answer in its simplest form.

c What is the scale factor of the enlargement that takes S to T?

d The largest angle is triangle S is 140°. What is the size of the largest angle in triangle T?

2 Rectangles R and S are similar.

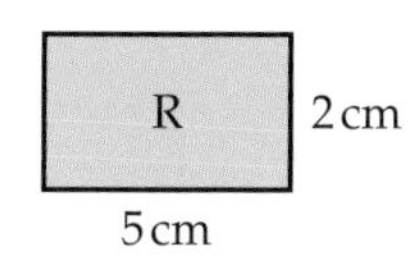

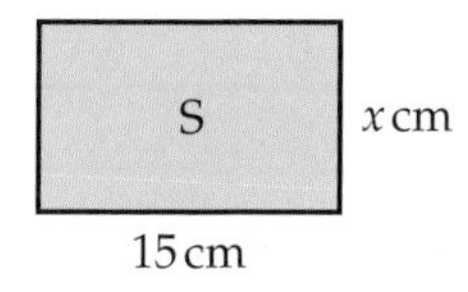

a Calculate the value of x.

b What is the ratio of the perimeters of the two rectangles?

B

AO2

3 Ali says that these two triangles are similar.
Beth says that the bigger triangle is an enlargement of the smaller triangle with a scale factor of 3.
Who is correct?
Give reasons for your answer.

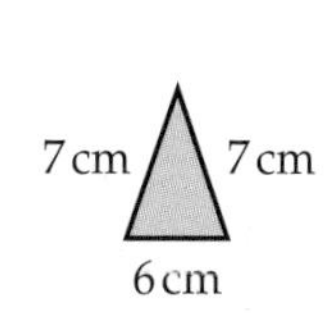

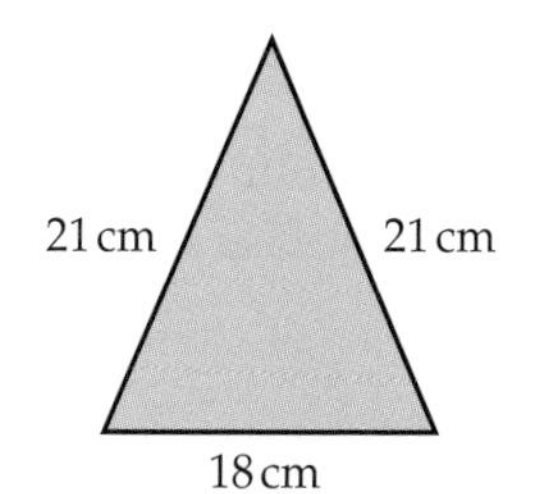

32 Non-linear graphs

Links to:
Middle Student Book
Ch32, pp.486–502

Key Points

Quadratic graphs D C

A quadratic function has a term in x^2. It may also have a term in x and a number. It does not have any terms with powers of x higher than 2.

The graph of a quadratic function is a curve, called a parabola, in the shape of a U.

All quadratic graphs are symmetrical about a line parallel to the y-axis.

The line of symmetry is given as 'x = a number'.

At the point where the curve turns, y has either a maximum or a minimum value.

Solving quadratic equations graphically C

Graphs can be used to solve quadratic equations. For example, solving $x^2 - 2x - 1 = 0$ means looking to see where the curve crosses the x-axis ($y = 0$).

Solving $x^2 - 2x - 1 = 6$ means looking to see where the curve crosses the line $y = 6$.

Graphs of cubic functions B

A cubic function has a term in x^3. It may also have a term in x^2 and x and a number. It does not have any terms with powers of x higher than 3.

You can solve a cubic equation by drawing its graph and finding where the graph crosses the x-axis.

32.1 Graphs of quadratic functions

D

1 a Copy and complete the table of values for $y = x^2 + 1$

x	−3	−2	−1	0	1	2	3
$y = x^2 + 1$	10	5				5	

Draw your y-axis going from −2 to +14 and your x-axis going from −3 to +3

b Draw the graph of $y = x^2 + 1$ for values of x from −3 to +3.

c Use your graph to estimate the value of $1.5^2 + 1$.

2 a Draw the graph of $y = x^2 - 1$ on the same axes you used in Q1.

b Describe the similarities and differences between the graphs of $y = x^2 + 1$ and $y = x^2 - 1$.

C

3 a Copy and complete the table of values for $y = x^2 + x$.

x	−3	−2	−1	0	1	2	3
x^2	9						9
$+x$	−3						+3
$y = x^2 + x$	6						12

b Draw the graph of $y = x^2 + x$ on the same axes you used in Q1.

c Describe the similarities and differences between the graphs of $y = x^2 + 1$ and $y = x^2 + x$.

4 Use your graphs in Q1, 2 and 3 to write down the line of symmetry for each of these graphs.

a $y = x^2 + 1$ b $y = x^2 - 1$ c $y = x^2 + x$

5 a Make a table of values and then draw the graph of $y = 2x^2 + 2x - 2$ for values of x from −3 to +2.

b Write down the minimum or maximum value of y.

c Write down the line of symmetry of the graph.

d Write down the coordinates of the x-intercept of the line of symmetry.

6 David wants to fence a section of a field for some calves.
He is investigating different sizes of rectangular enclosures.
The graph shows how the area (A cm²) of the enclosure changes as the value of x varies.

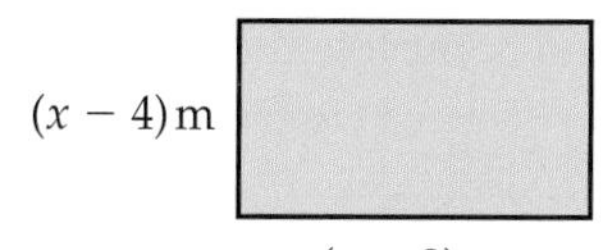

C

a The area of the enclosure needs to be greater than 25 m².
What is the minimum value of x that David can consider?

b What is the area of the enclosure when $x = 2$?

c What is the area of the enclosure when $x = 5$?

d Look at your answers to parts **b** and **c**.
Which of these values of x would be possible in real life?
Give a reason for your answer.

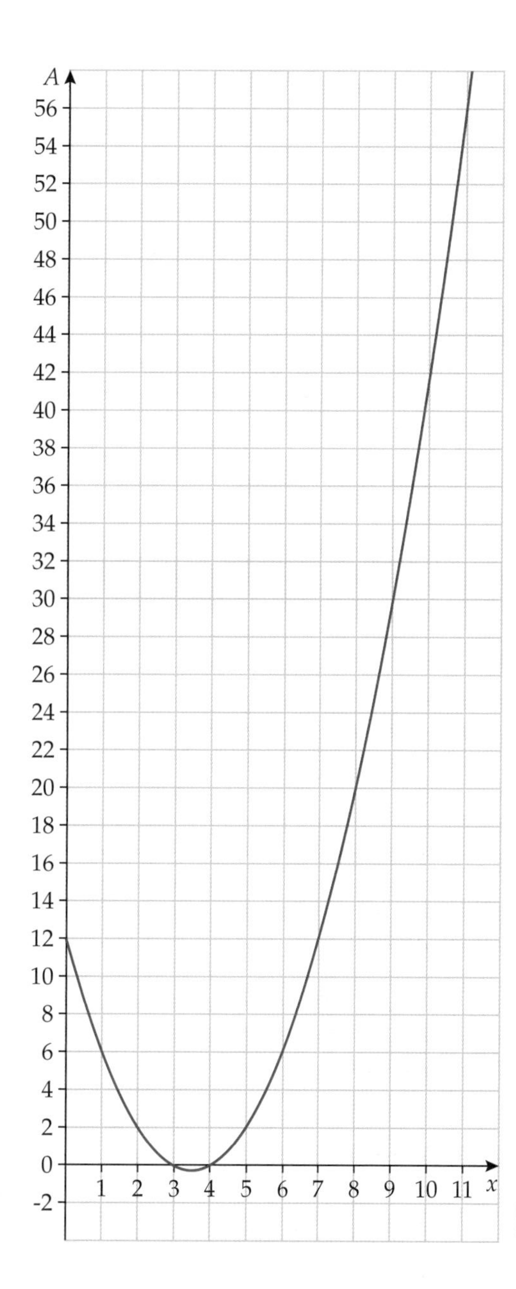

AO2

32.2 Solving quadratic equations graphically

1 **a** Make a table of values and then draw the graph of $y = x^2 - 3x + 1$ for values of x from -3 to $+3$.

b An approximate solution of the equation $x^2 - 3x + 1 = 0$ is $x = 0.4$
 i Explain how you can find this from your graph.
 ii Use your graph to find another solution to this equation.

Draw your y-axis going from -2 to 20 and your x-axis going from -3 to $+3$.

C

C

2 a Copy and complete the table of values for $y = 2x^2 - 5$.

x	−3	−2	−1	0	1	2	3
y	13	3					13

b Draw the graph of $y = 2x^2 - 5$.

c Use your graph to find the solutions of the equation $2x^2 - 5 = 0$.

d Draw the line $y = 8$ on your graph.
Write down the coordinates of the points where the line and the curve cross.

e Show that the solutions to the quadratic equation $2x^2 - 13 = 0$ can be found at this point.

> Draw your y-axis going from −5 to 15 and your x-axis going from −3 to +3.

A02

C

3 a Draw the graph of $y = x^2 - 5x + 2$ for values of x from −1 to +5.

b Use your graph to solve
 i $x^2 - 5x + 2 = 0$
 ii $x^2 - 5x + 2 = -3$

c Can the quadratic $x^2 - 5x + 2 = -5$ be solved? Explain your answer.

> Draw your y-axis going from −10 to +10 and your x-axis going from −1 to +5.

A03

32.3 Graphs of cubic functions

B

1 a Copy and complete the table of values for the function $y = x^3 - 4x + 4$.

x	−3	−2	−1	0	1	2	3
x^3	−27			0		8	
$-4x$	12			0		−8	
$+4$	+4			+4		+4	
$y = x^3 - 4x + 4$	−11			4		4	

b Draw the graph of $y = x^3 - 4x + 4$ for $-3 \leqslant x \leqslant 3$.

c Use your graph to find the solution to $x^3 - 4x + 4 = 0$.

2 a Draw the graph of $y = x^3 - 3x^2 - x$ for $-2 \leqslant x \leqslant 4$.

b Use your graph to find the solution to $x^3 - 3x^2 - x = 0$.

3 Without drawing a graph, write down the solutions to $(x - 1)(x - 3)(x + 5) = 0$.

4 The graphs of four functions are shown below.

A
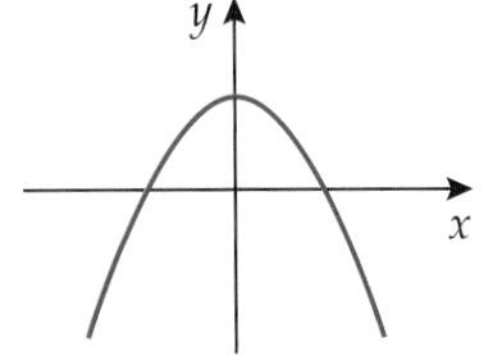

B
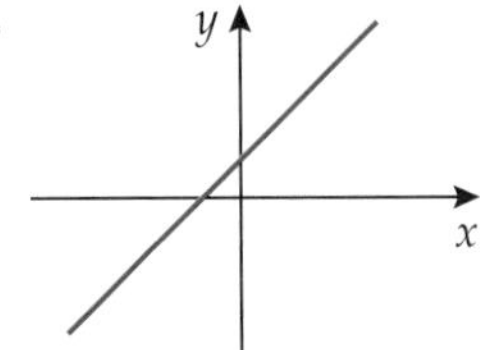

C
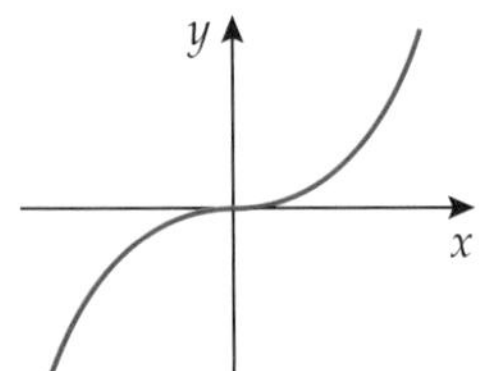

D
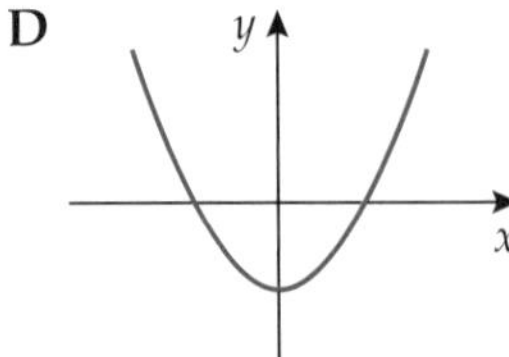

a Match each graph to the correct function.

$y = x^3$ $\quad y = x^2 - 4$ $\quad y = 4x + 4$ $\quad y = -x^3$ $\quad y = -x^2 + 4$

b Sketch the graph of the function that is not used.

Links to:
Middle Student Book
Ch33, pp.503–519

Key Points

Constructions C

Constructions must be drawn using **only** a straight edge (ruler) and a pair of compasses.

Leave all construction lines and arcs on the diagram as evidence you have used the correct method.

Perpendicular at a point on a line

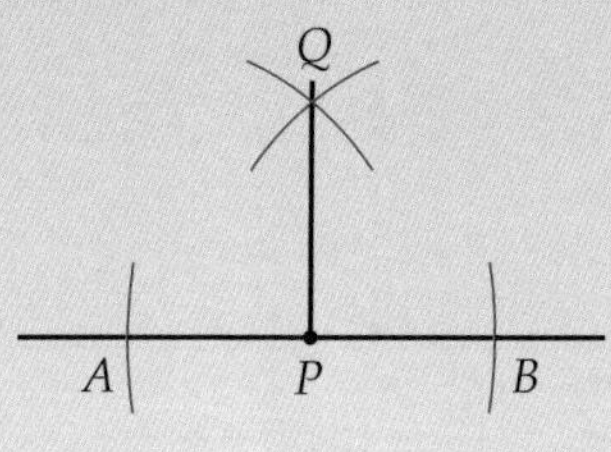

Perpendicular from a point to a line

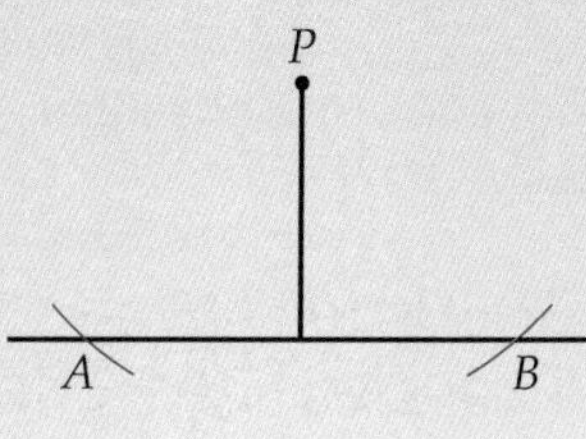

Perpendicular bisector of a line segment

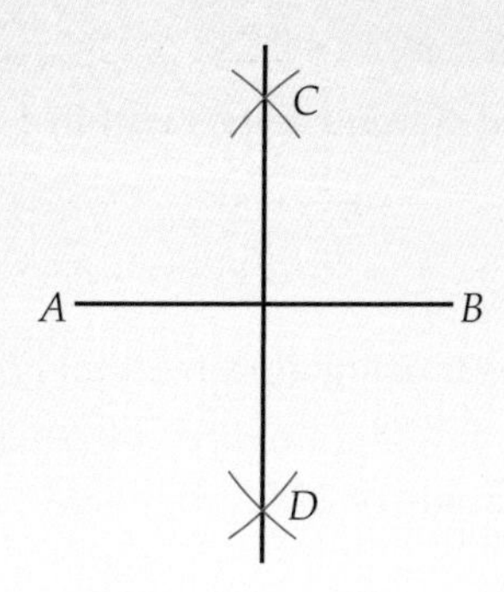

Angle of 60°

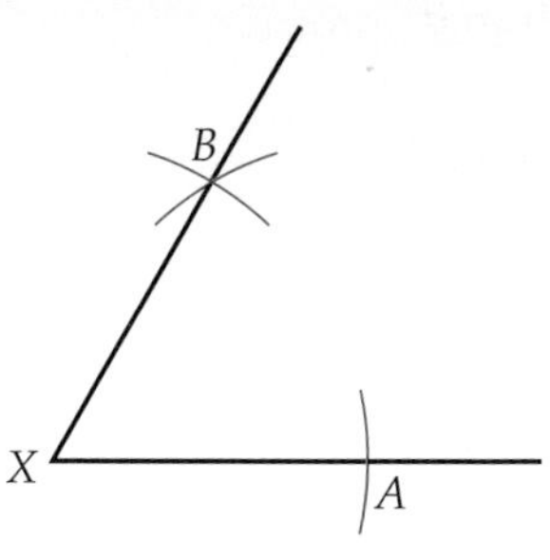

The bisector of an angle

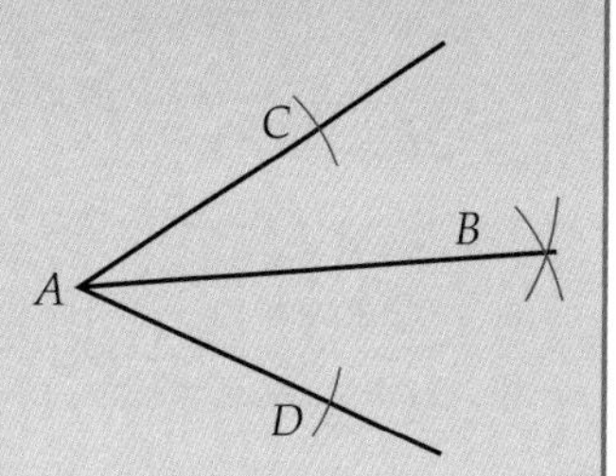

Locus C

The locus of points that are the same distance from a fixed point is a circle.

The locus of points that are the same distance from a fixed line is two parallel lines, one each side of the given line.

The locus of points that are the same distance from a fixed line segment AB is a 'racetrack' shape. The shape has two lines parallel to AB and two semicircular ends.

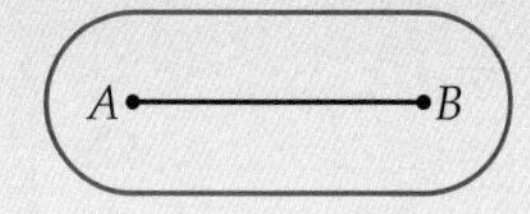

The locus of points that are equidistant from two fixed points is the perpendicular bisector of the line segment joining the two points.

The locus of points that are equidistant from two fixed lines is the bisector of the angle formed by the lines.

33.1 Constructions

1 a Draw a line segment 6 cm in length.

b Construct the perpendicular bisector of the line segment.

c Label the mid-point of your line segment M.

C

C

2 a Draw a circle of radius 5 cm.

b Draw a triangle inside your circle so that all three vertices of the triangle touch the circumference of the circle. For example:

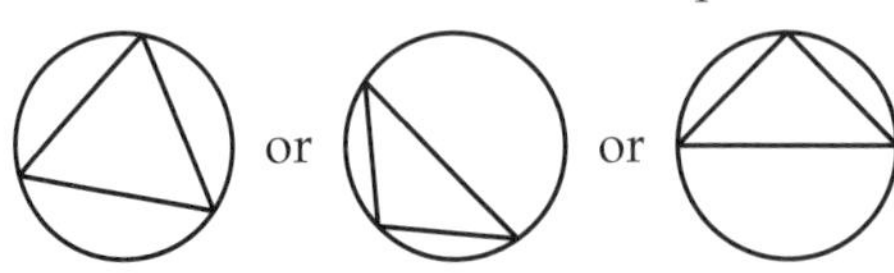

c Construct the perpendicular bisector of each side of your triangle.

d Write down what you notice.

3 Make an accurate copy of this diagram.
You will need to construct the perpendicular from A to the line.

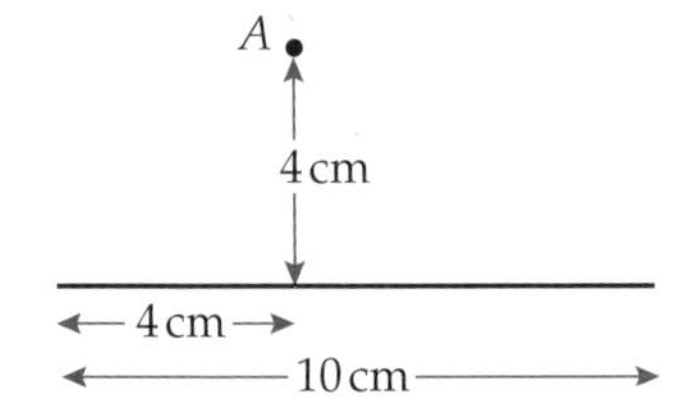

4 Construct a square on plain paper with only compasses and a straight edge (not a ruler). You must show all your construction lines.

C

5 a Accurately draw a rectangle $ABCD$ with $AB = 10$ cm and $AD = 5$ cm.

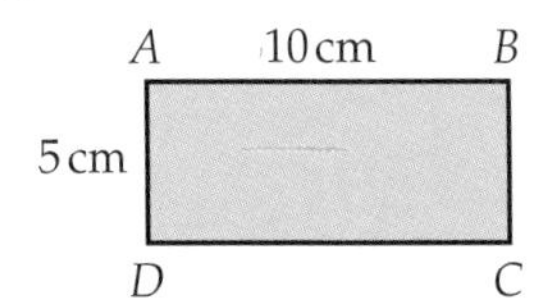

b Construct angle bisectors of angles ADC and ABC.

c Write down the mathematical names of the two different shapes within the rectangle, formed by the angle bisectors.

AO2

d Work out the areas of each of these shapes.

C

6 a Gillian says she can draw an angle of 22.5° without using a protractor. Explain how she can do this.

b Using only a ruler and compasses, construct an angle of 22.5°.

7 a Adam says he can draw an angle of 60° without using a protractor. Explain how he can do this.

b Using only a ruler and compasses, construct an angle of 60°.

8 Make a list of all the acute angles that you could accurately draw using only a ruler and compasses.

33.2 Locus

C

1 The line AB is 7 cm long.
Make a copy of the line AB.
Draw the locus of the points that are exactly 3 cm from AB.

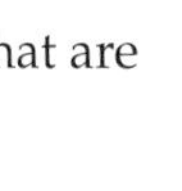

2 a Use a ruler to make a copy of this line.

b Draw the locus of points that are 2 cm from the line.

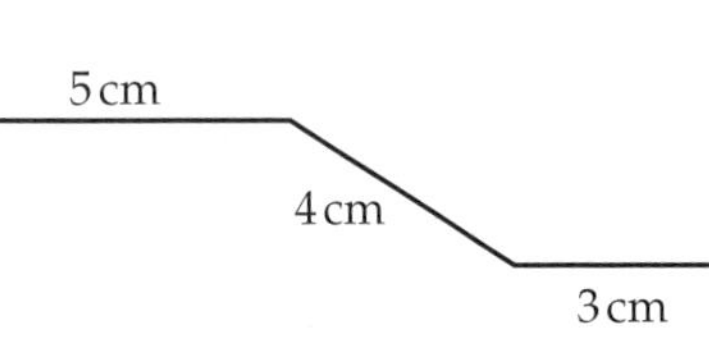

3 A chimpanzee enclosure at a zoo has bars wide enough for the chimps to put their arms through.
A chimpanzee's arm is approximately 1 m long.
Using a scale of 3 cm to 1 m, show how far a chimp can reach outside of this section of the enclosure.

4 *ABDC* is a rectangular garden.
There is a fence around the perimeter of the garden.
A goat is tethered by an 8 m rope to corner *C*.
There is a shed along the fence *AC*.

a Make a scale diagram of the garden using a scale of 1 cm to 2 m.

b Show on the diagram all the possible positions of the goat if the rope remains tight.

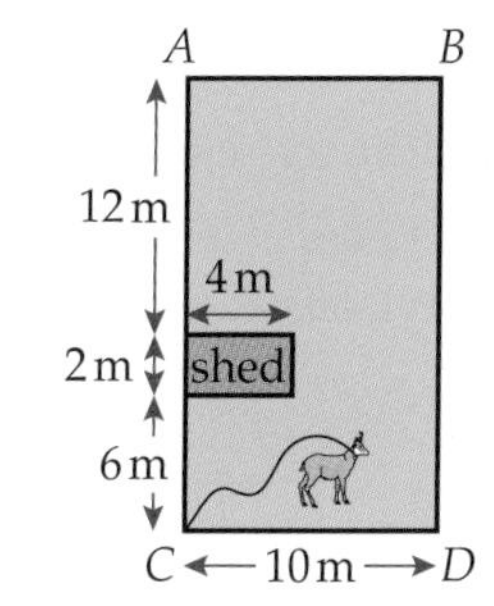

5 a Copy this diagram.

b Construct the angle bisector of angle *CAB*.

c Shade the region inside the triangle of the points that are closer to *AC* than to *AB*.

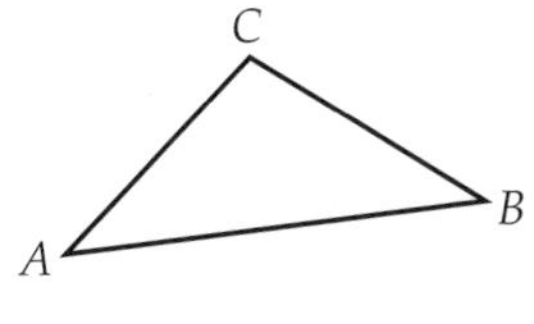

6 Two transmitters *A* and *B* are 80 km apart.
Transmitter *A* can broadcast a distance of 60 km.
Transmitter *B* can broadcast a distance of 40 km.
Draw a scale diagram to show clearly the region covered by both transmitters.

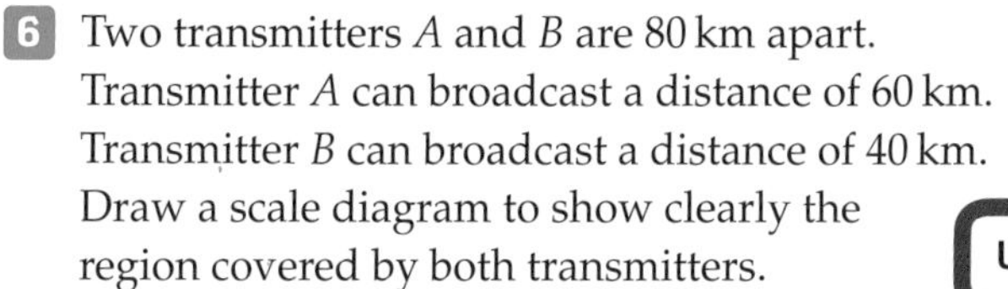

Use a scale of 1 cm to 10 km.

7 *ABCD* is a rectangular lawn.

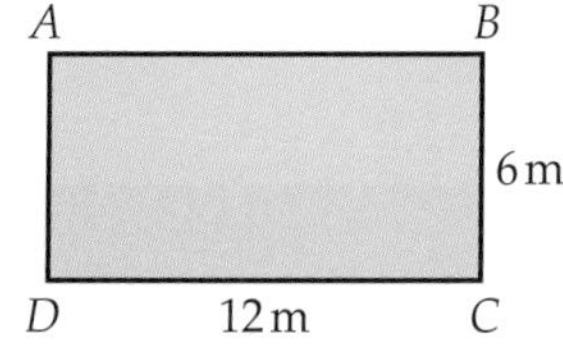

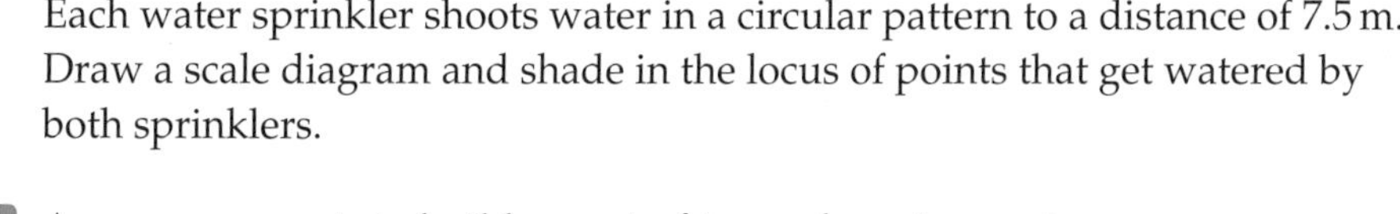

Water sprinklers are placed at the mid-points of side *AD* and side *BC*.
Each water sprinkler shoots water in a circular pattern to a distance of 7.5 m.
Draw a scale diagram and shade in the locus of points that get watered by both sprinklers.

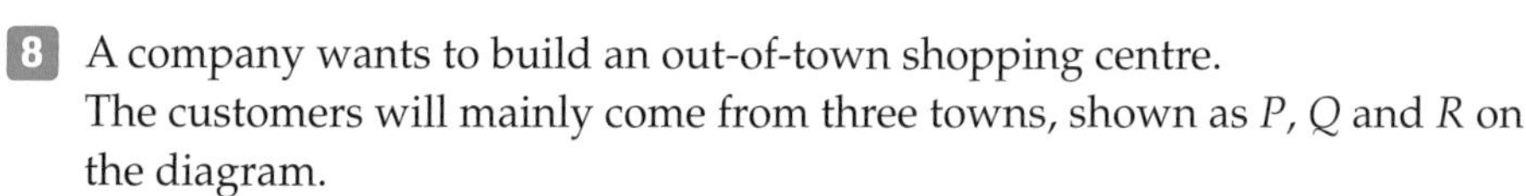

8 A company wants to build an out-of-town shopping centre.
The customers will mainly come from three towns, shown as *P*, *Q* and *R* on the diagram.

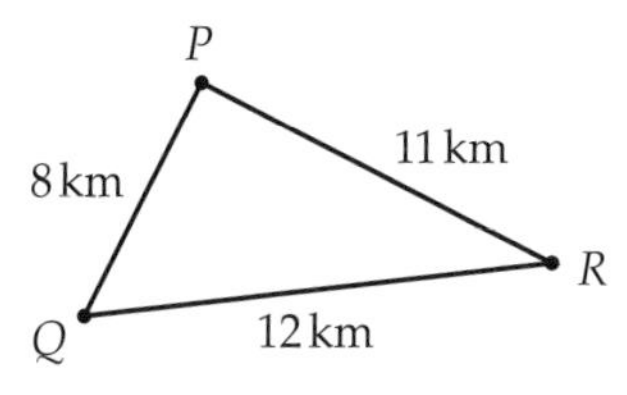

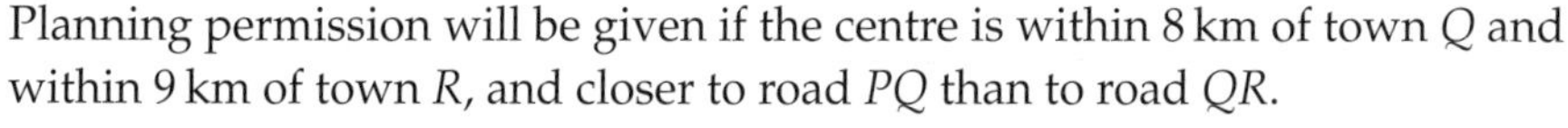

Planning permission will be given if the centre is within 8 km of town *Q* and within 9 km of town *R*, and closer to road *PQ* than to road *QR*.

a Make a copy of the diagram using a scale of 1 cm to 1 km.

b Shade the region in which the company can build the shopping centre.

C

AO3

34 Pythagoras' theorem

Links to:
Middle Student Book
Ch34, pp.520–532

Key Points

Pythagoras' theorem **C**

In a right-angled triangle the longest side is called the hypotenuse.

The hypotenuse is always opposite the right angle.

For a right-angled triangle with sides of length a, b and c, where c is the hypotenuse, Pythagoras' theorem states that $a^2 + b^2 = c^2$.

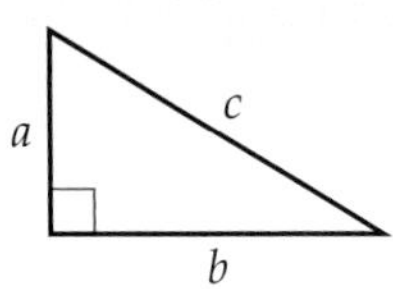

You can calculate the lengths of the shorter side of a right-angled triangle using these equations.

- $a^2 = c^2 - b^2$
- $b^2 = c^2 - a^2$

Length of a line segment **C**

Pythagoras' theorem can be used to find the length of a diagonal line AB, given the coordinates of A and B.

34.1 Pythagoras' theorem

C

1 Write down the letter that represents the hypotenuse in this triangle.

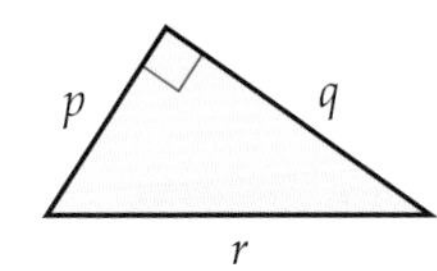

2 For this triangle

a write down the letter that represents the hypotenuse

b write down the formula for Pythagoras' theorem.

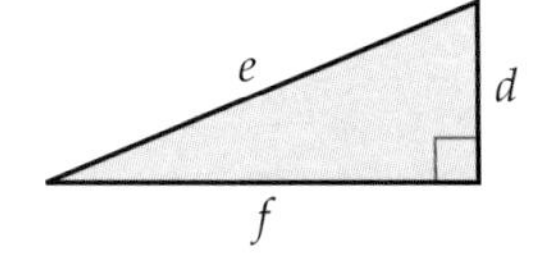

34.2 Finding the hypotenuse

C

1 Calculate the length of the hypotenuse in this triangle.

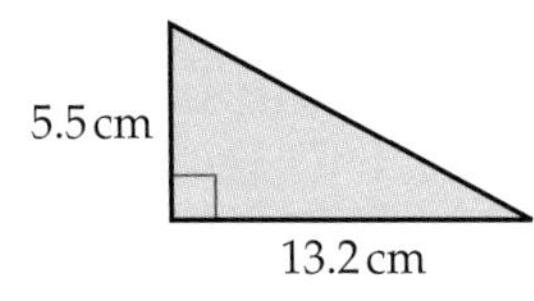

2 A rectangular piece of A5 paper measures 210 mm by 148 mm.
What is the length of the diagonal on a sheet of A5 paper?
Give your answer to the nearest millimetre.

C **AO2**

3 Mr Jones builds a lean-to shed on the side of his house.
One wall of the shed is 2.4 m high and the other is 1.8 m high.
The horizontal distance between the walls is 2.2 m.
Work out the length of the roof.
Give your answer correct to the nearest centimetre.

> **Draw a right-angled triangle and write on the two measurements that you know. The height of the triangle is *not* 2.4 m.**

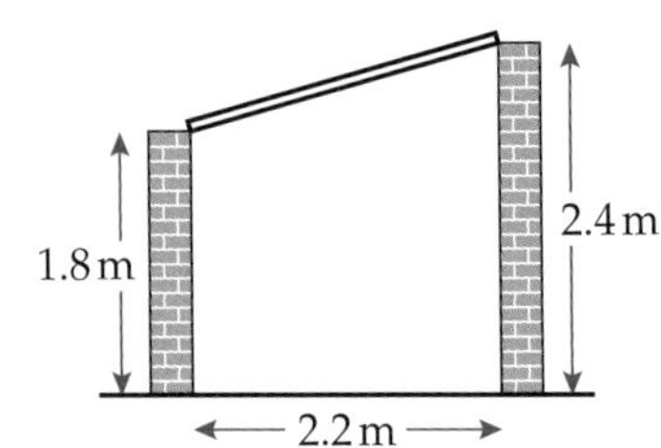

4 The diagram shows a sketch of part of Iqbal's garden.
He is going to have wooden decking in one area and lawn in the rest.
The carpenter needs to know the length of the diagonal join.
Work out the length of this join.
Give your answer correct to the nearest centimetre.

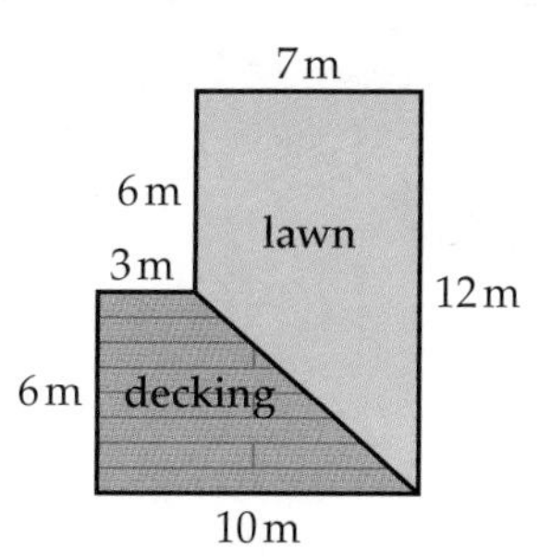

C
AO3

34.3 Finding a shorter side

1 Calculate the length of the side marked x in this triangle.

> **Remember to use a rearranged version of Pythagoras' theorem as you are finding one of the shorter sides.**

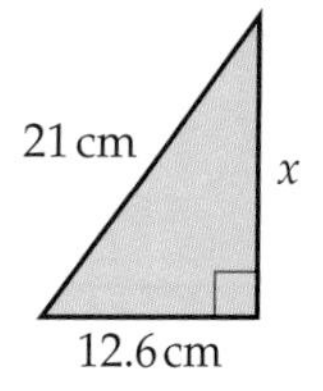

C

2 The diagram shows the width and length of a diagonal of a sheet of A4 paper.
Calculate the area of the sheet of A4 paper.
Give your answer correct to 1 d.p.

> **Remember, area of rectangle = base × height.**

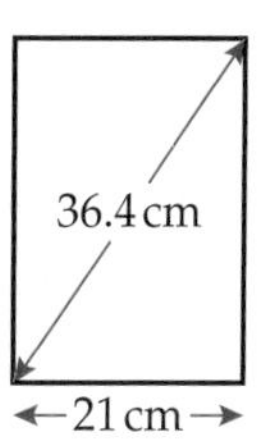

C

3 For safety reasons a 3.5 m ladder must be placed between 0.9 m and 1.25 m from the base of a wall.
What is the difference in the heights that the ladder can reach up the wall?
Give your answer to the nearest centimetre.

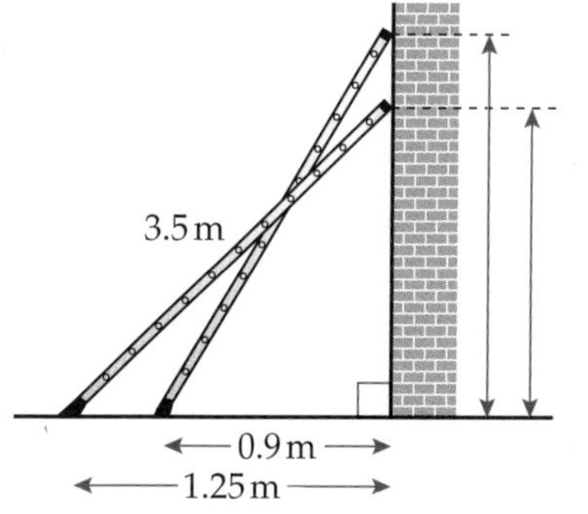

AO2

4 A plane flies 180 km south-west, then 135 km north-west.
How far is the plane from its starting point?

> **Draw a sketch to help.**

C
AO3

34.4 Calculating the length of a line segment

1 Calculate the length of each of these lines.
Give your answer correct to 1 d.p. where appropriate.

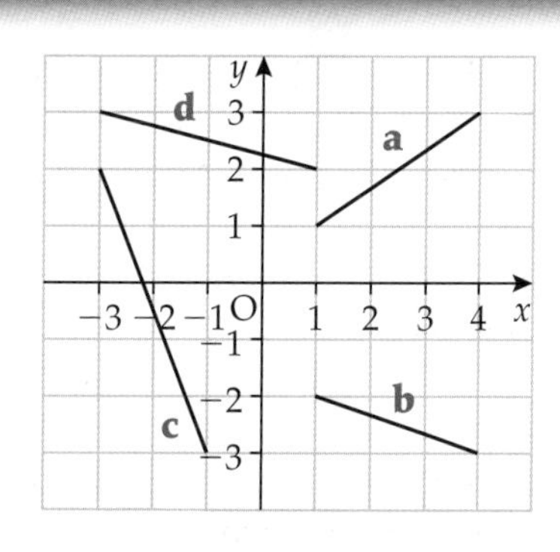

C

2 Calculate the length of the line segment from $(-2, 1)$ to $(4, 9)$.

> **Draw a sketch to help.**

C
AO2

35 Trigonometry

Links to:
Middle Student Book Ch35, pp.533–546

Key Points

Trigonometry

B

Trigonometric functions are used to find lengths or angles in right-angled triangles.

The basic trigonometric functions are given by the following equations.

$\sin x = \dfrac{\text{opposite}}{\text{hypotenuse}} = \dfrac{O}{H}$

$\cos x = \dfrac{\text{adjacent}}{\text{hypotenuse}} = \dfrac{A}{H}$

$\tan x = \dfrac{\text{opposite}}{\text{adjacent}} = \dfrac{O}{A}$

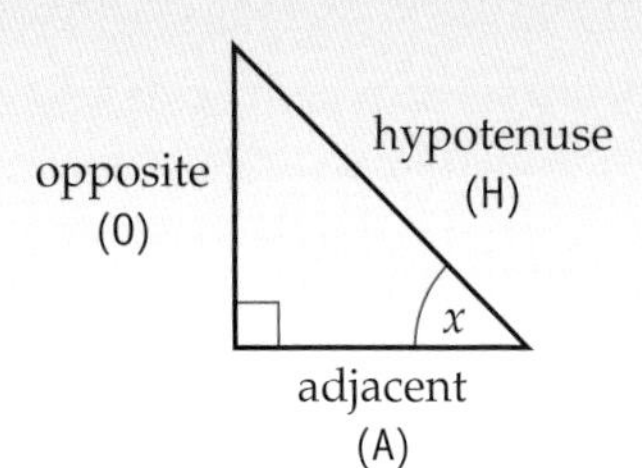

35.1 Trigonometry – the ratios of sine, cosine and tangent

B

1 Sketch this triangle and label the sides H, O and A, where x is the angle to be found.

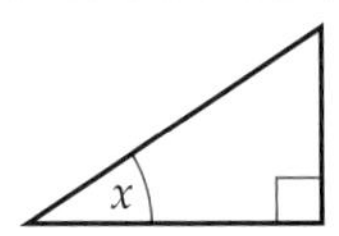

2 Use your calculator to work out the values of the following trigonometric functions.

Give your answers correct to 5 d.p.

a $\sin 44°$ **b** $\cos 44°$ **c** $\tan 44°$

d $\sin 144°$ **e** $\cos 144°$ **f** $\tan 144°$

3 Use your calculator to find the value of each angle.

Give your answers correct to 1 d.p.

a $\sin x° = 0.5821$ **b** $\cos x° = 0.9816$ **c** $\tan x° = 2.475$

35.2 Finding lengths using trigonometry

Calculate the length x marked on each of the following diagrams.
Give your answer correct to 1 d.p. where appropriate.
All lengths are in centimetres.

B

1

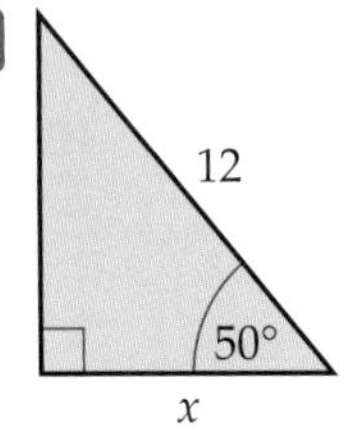

2

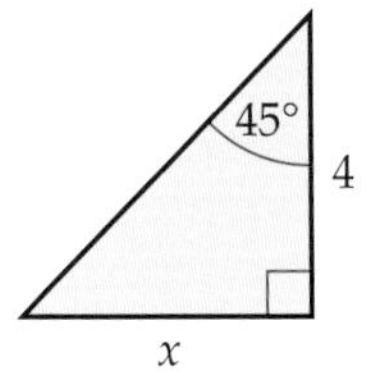

3

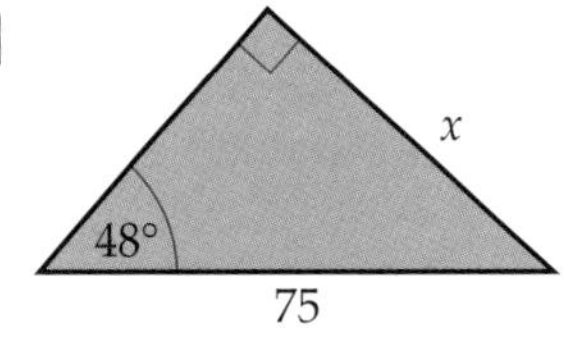

4

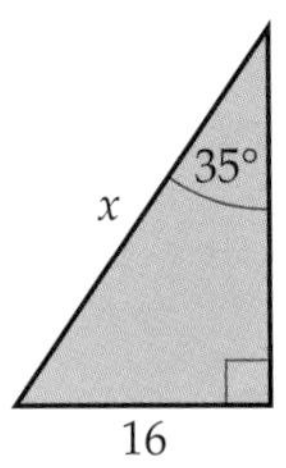

5

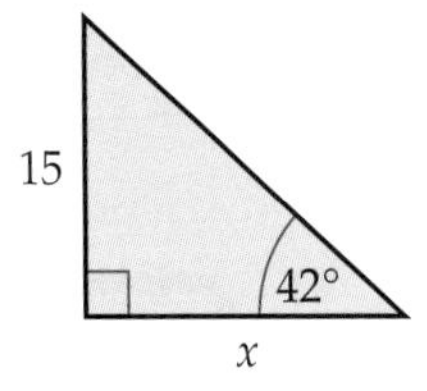

6

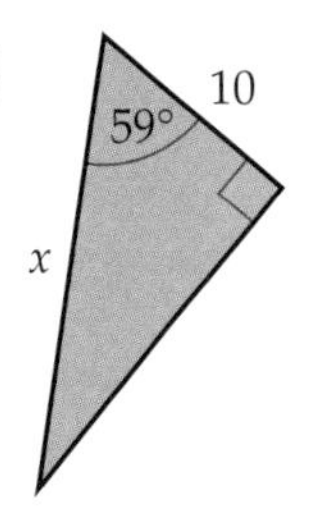

35.3 Finding angles using trigonometry

1 Calculate the size of angle x in each of the following triangles.
Give your answers to 3 s.f. where appropriate.

a 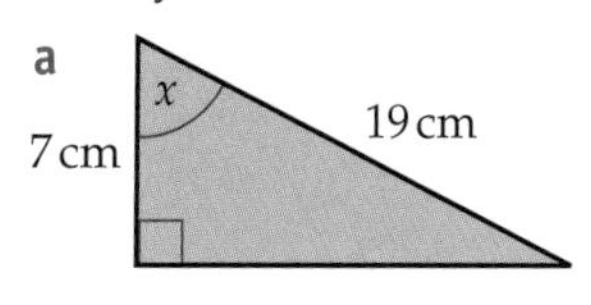

b 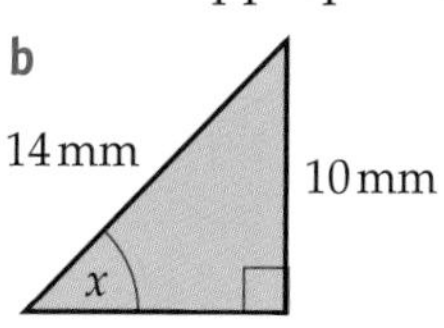

c 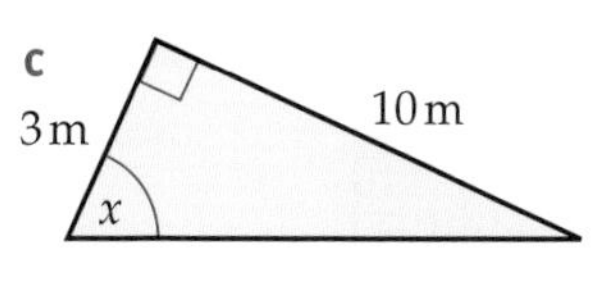

2 This is part of Tamsin's homework.

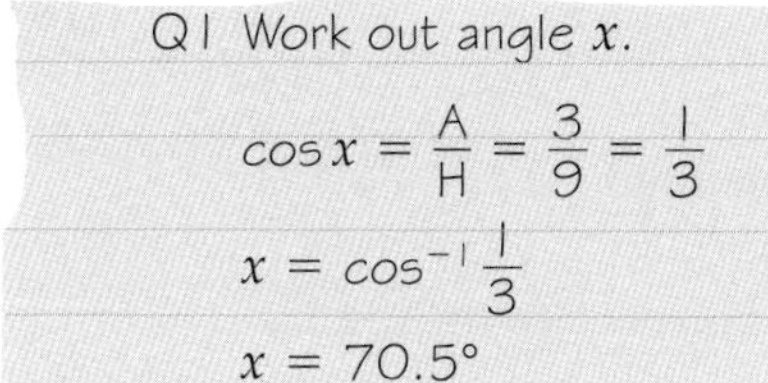

Q1 Work out angle x.

$\cos x = \frac{A}{H} = \frac{3}{9} = \frac{1}{3}$

$x = \cos^{-1} \frac{1}{3}$

$x = 70.5°$

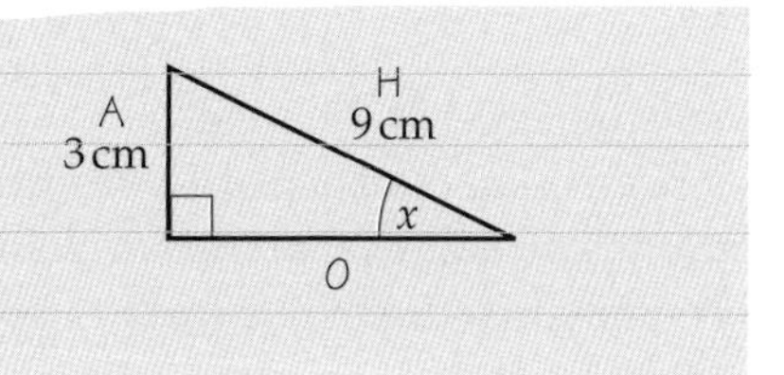

a What mistake has Tamsin made?

b Work out the correct answer for her.

B

AO2

35.4 Applications of trigonometry and Pythagoras' theorem

1 The positions of three oil rigs A, B and C are shown on the diagram.
A is 9 miles due south of B.
C is 7 miles due east of B.
Work out the bearing of C from A.
Give your answer to the nearest degree.

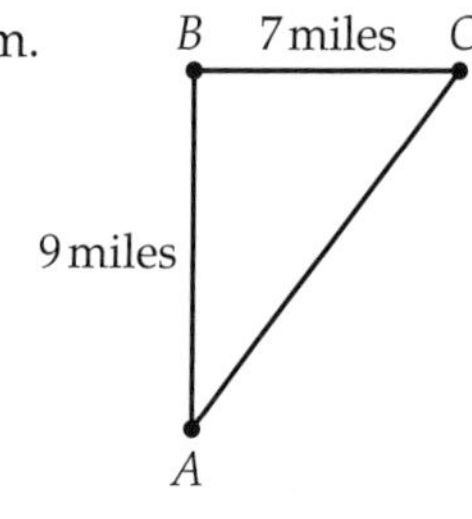

B

2 The positions of three oil rigs X, Y and Z are shown on the diagram.
X is 18 km due west of Z.
Y is due north of Z.
The bearing of Y from X is 050°.
Work out the distance of Y from Z.
Give your answer correct to 2 d.p.

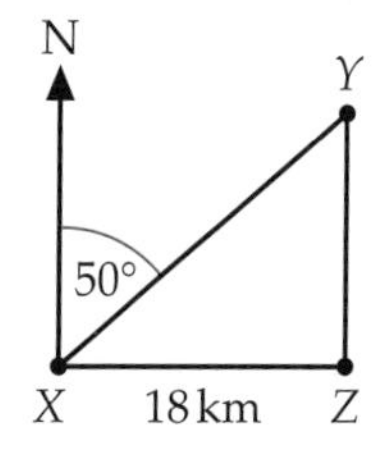

B

3 A 'death-slide' is a steel cable that joins two platforms.
The diagram shows the two platforms, one 27 m high and the other 5 m high.
The horizontal distance between the platforms is 80 m.
Work out the angle the cable makes with the horizontal.
Give your answer correct to the nearest 0.1°.

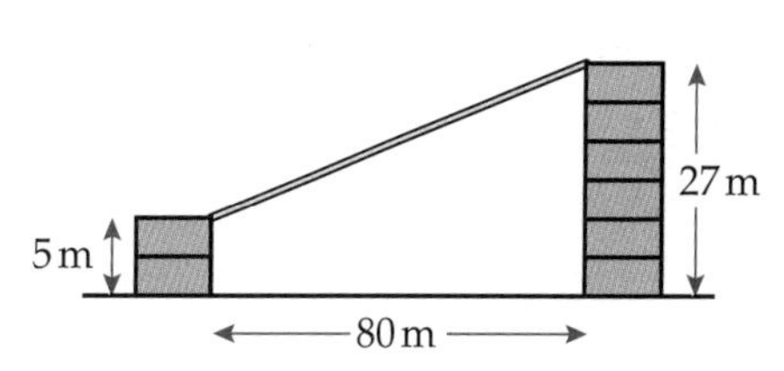

AO2

4 The diagram shows a right-angled triangle in a circle with centre O.
The radius of the circle is 10 cm.
Calculate the area of the triangle.
Give your answer correct to 1 d.p.

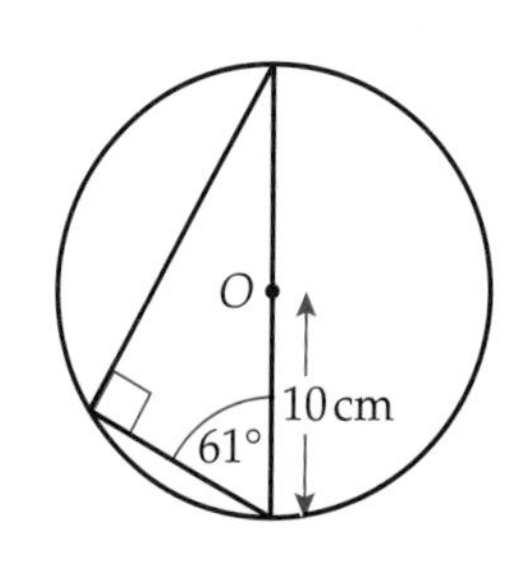

B

5 A new house needs a driveway to be built from its car parking space to the road.
The diagrams show the plan view and the side view of the house and road.

Plan view

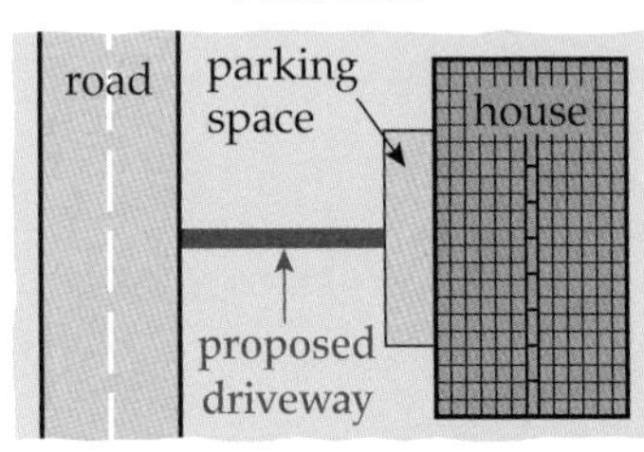

Side view

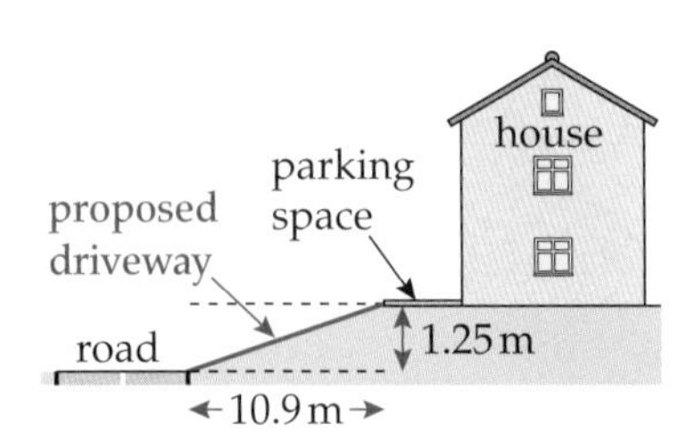

The property is 1.25 m above the level of the road.
The car parking space is a horizontal distance of 10.9 m from the road.
For a driveway the maximum safe angle from the horizontal is 6°.
Can the driveway be built safely in a direct line from the parking space to the road?

AO3

35.5 Angles of elevation and depression

B

1 Steven is 45 m from the base of a cliff.
The angle of elevation to the top of the cliff is 42°.
How high is the cliff? Give your answer to the nearest metre.

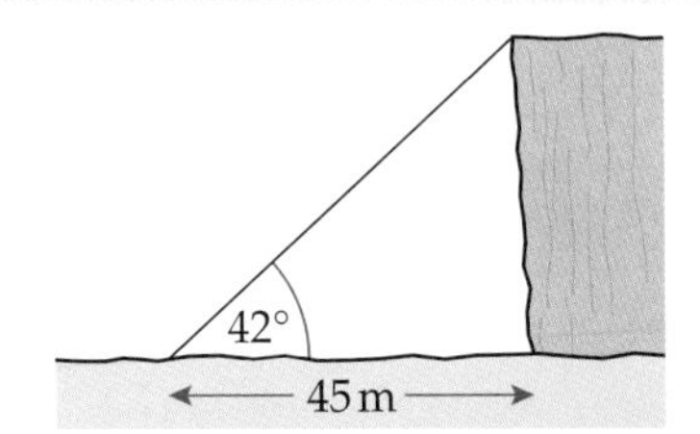

2 Ted is standing on the beach looking out to sea at a lighthouse.
The lighthouse is 18 km from the beach.
The angle of elevation of the top of the lighthouse is 0.1°.
How tall is the lighthouse?

3 From the top of a building, the angle of depression of an ice cream van is 55°.
The ice cream van is 60 m from the base of the building.
How high is the building?

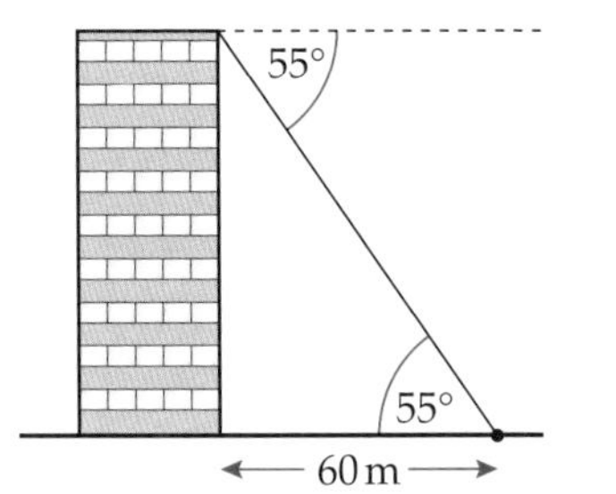

B

4 Shelley is standing 10 m from the foot of Nelson's Column in London.
Nelson's Column is a granite column with a statue on the top.
The angle of elevation to the top of the column is 77.7°.
The angle of elevation to the top of the statue is 79.0°.
Calculate the height of the statue.
Give your answer correct to 3 s.f.

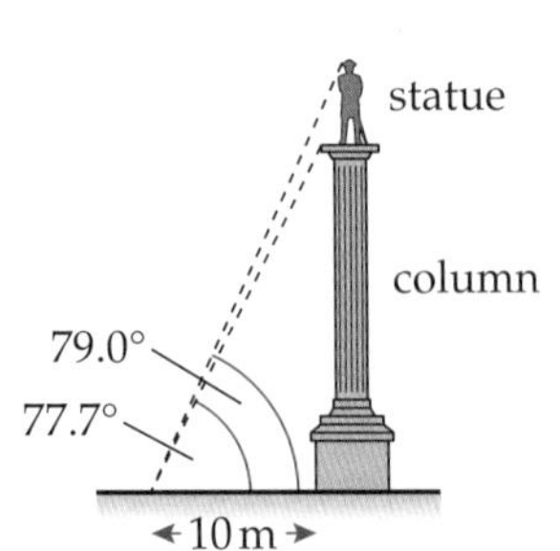

5 From the top of a 50 m cliff two boats can be seen in line.
The angles of depression of the two boats are 20° and 34° respectively.
Calculate the distance between the two boats. Give your answer to the nearest metre.

AO3

36 Circle theorems

Key Points

Circle properties

The perpendicular from the centre of a circle to a chord bisects the chord.

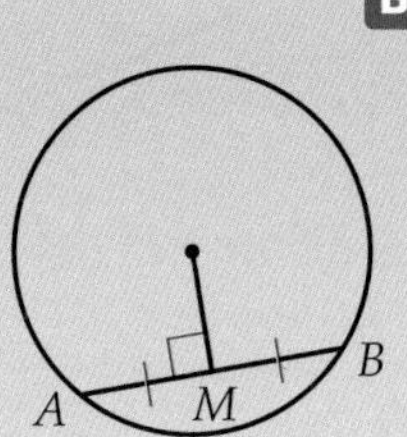

Tangents drawn to a circle from an external point are equal in length.

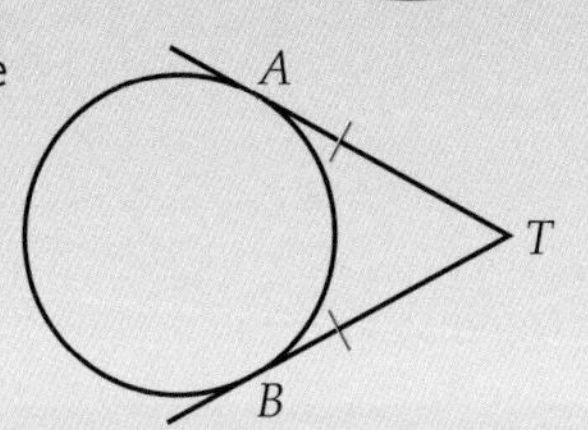

The angle between a tangent and radius is 90°.

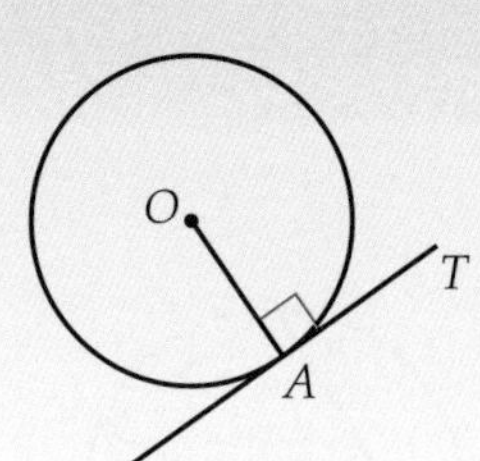

B The angle in a semicircle is a right angle.

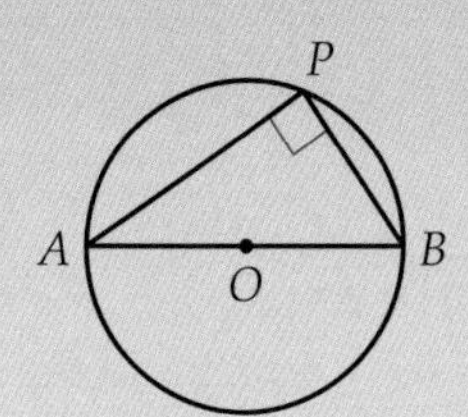

Angles subtended by the same arc are equal.

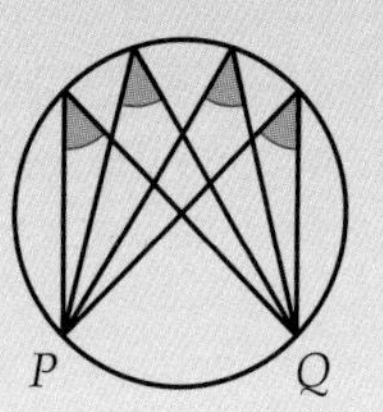

Opposite angles in a cyclic quadrilateral are supplementary.

$b + d = 180°$

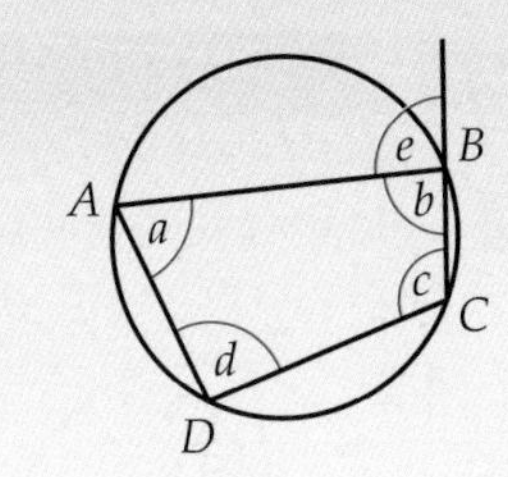

The exterior angle of a cyclic quadrilateral is equal to the opposite interior angle.

$e = d$

Circle theorems

B The angle subtended by an arc at the centre of a circle is twice the angle that it subtends at the circumference.

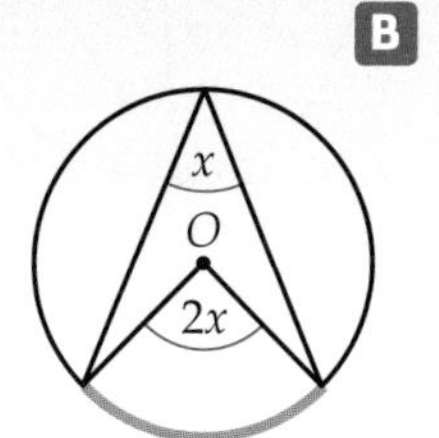

36.1 Circle properties

B

1 In this diagram, the length of the chord AB is 20 cm.
The mid-point of the chord is 2 cm from the centre of the circle.
Calculate the length of the radius of the circle.
Write your answer correct to 3 s.f.
Write down any circle properties that you use.

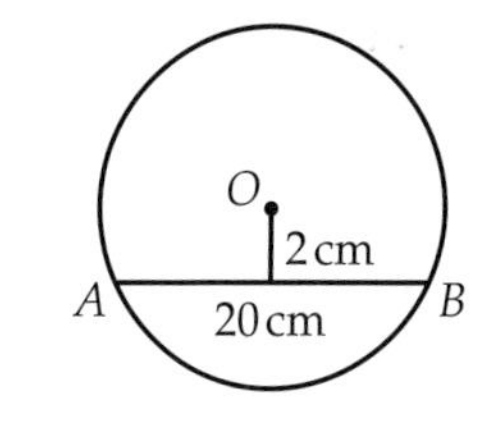

2 In this diagram, the circle with centre O has a radius of 8 cm.
The length of the chord AB is 9.6 cm.
Using Pythagoras' theorem, find the length of OM, where M is the mid-point of AB.

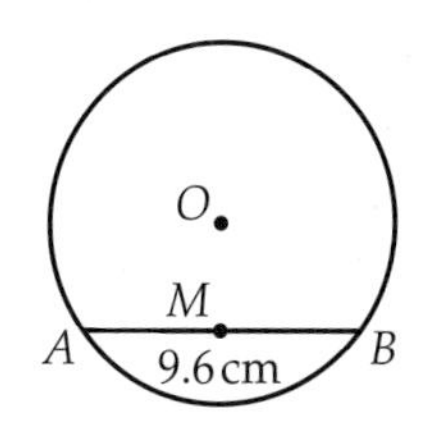

3 Work out the size of angle x in this diagram.
Show all your workings.

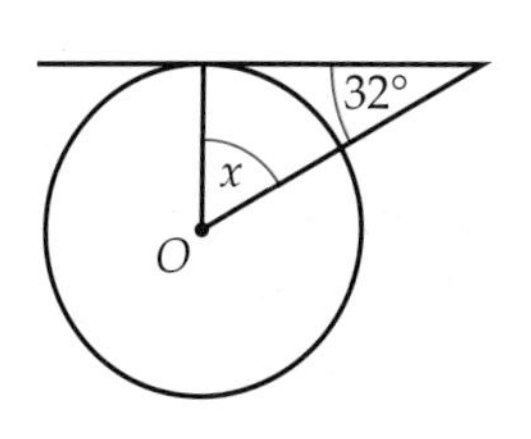

4 Calculate the size of angle x in this diagram.
Show all your working.

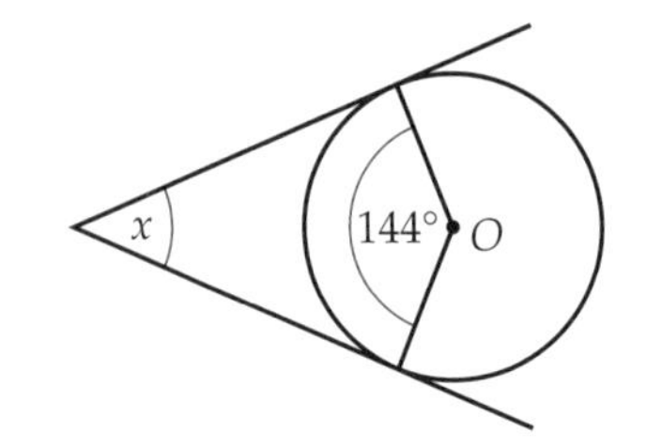

B

5 In this diagram, the sides of the triangle PQR are tangents to the circle.
$QU = 10$ cm and $PS = 8$ cm.

a Write down the length of QT.

b The perimeter of the triangle is 50 cm.
Calculate the length of SR.

AO2

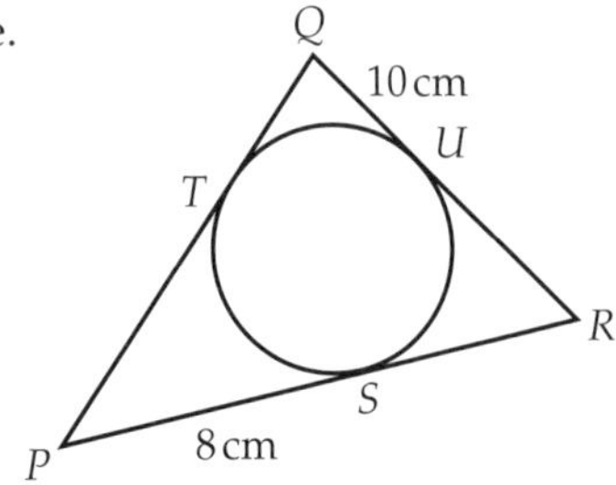

36.2 Circle theorems

B

1 Calculate the sizes of the angles marked with letters.
Explain each step in your reasoning.

a

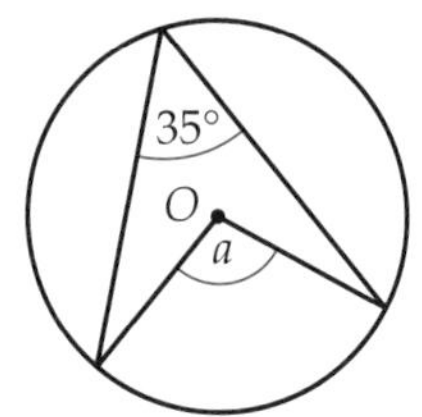

b

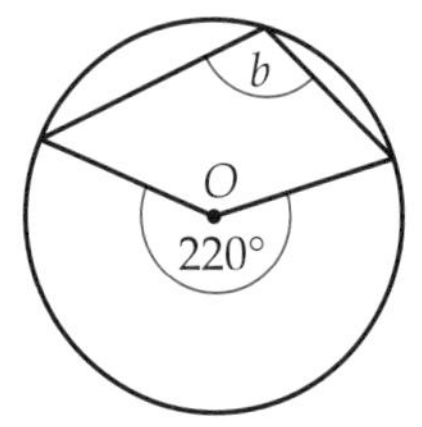

c

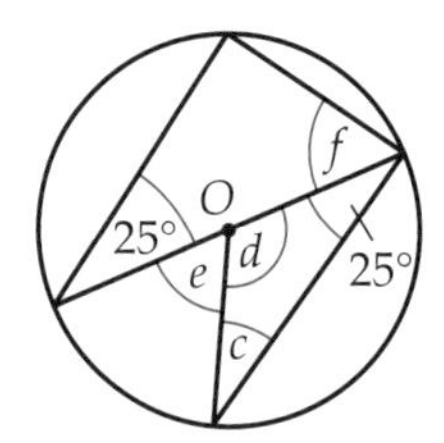

B

2 Calculate the size of angle x in this diagram.
Show every step in your working.

AO2

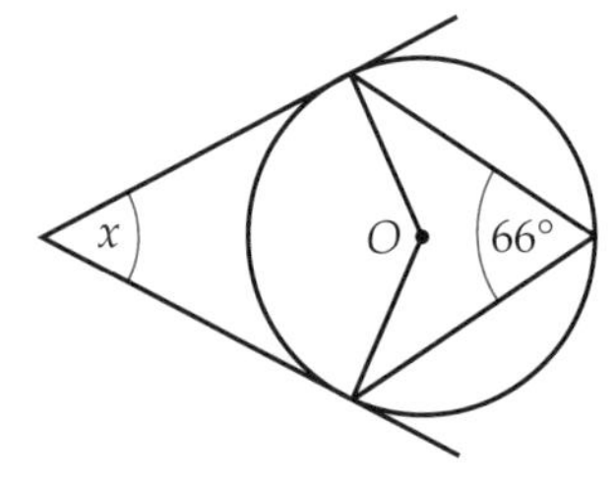

B

3 Calculate the sizes of the angles marked with letters. Explain each step in your reasoning.

a

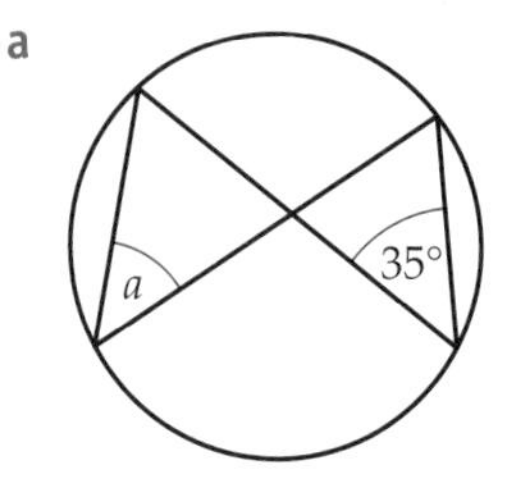

b

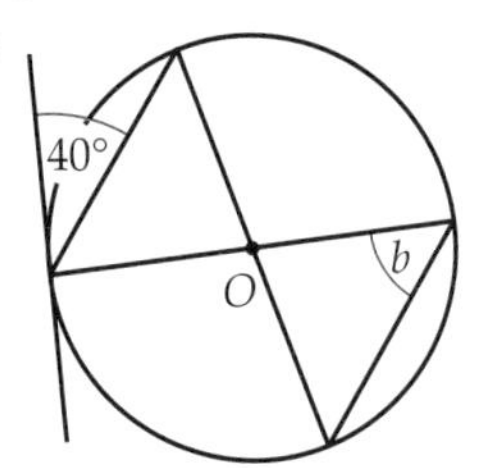

c

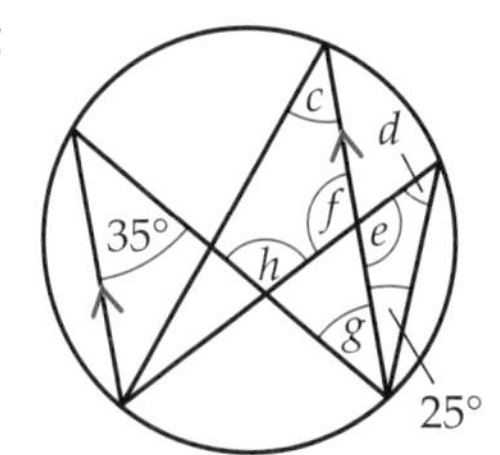

4 Calculate the sizes of the angles marked with letters. Explain each step in your reasoning.

a

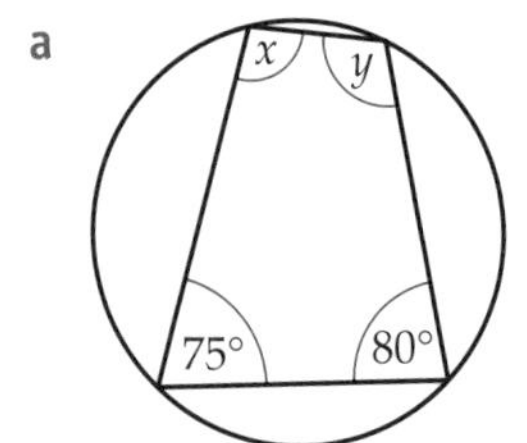

b

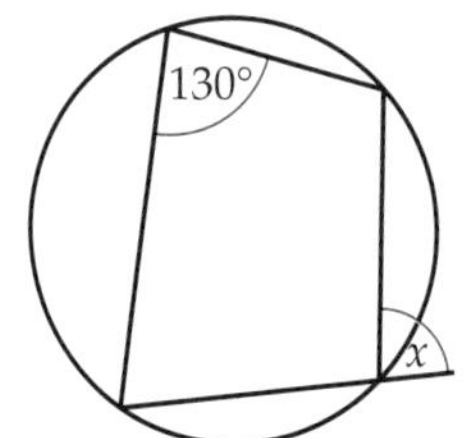

c

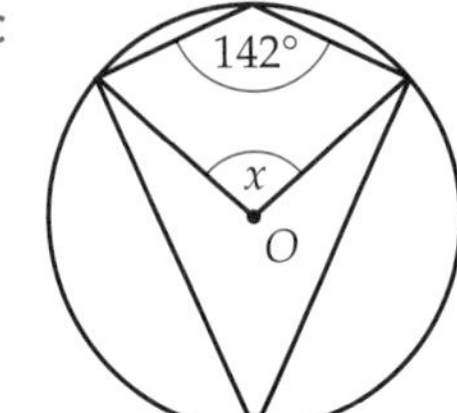

d

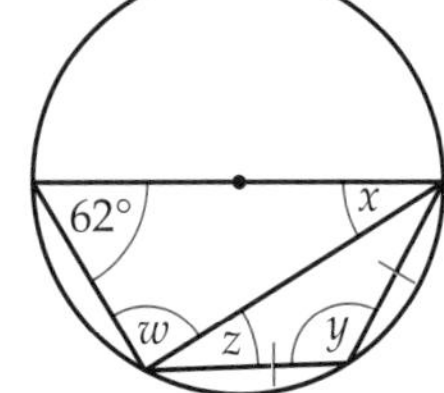

EXAM PRACTICE PAPERS

We have included Higher Practice Papers for Modular Units 1 and 2, along with Foundation for Unit 3.
We have included Higher Linear Practice Papers 1 and 2.

Unit 1 Higher Statistics and Number — Calculator allowed

1 Indie has these two spinners.

She spins the spinners at the same time.

She subtracts the smaller number from the larger number to give her a score.

The table shows her possible scores.

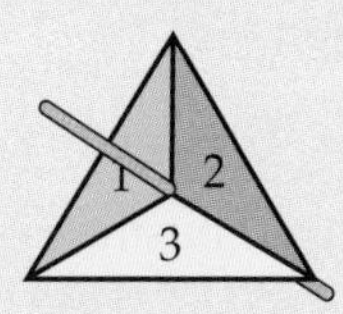

–	1	2	3	4	5
1	0	1	2	3	4
2	1	0	1	2	3
3	2	1	0	1	2

a Work out the probability Indie gets a score of 0. **(1 mark)** | D

b Work out the probability that her score is an odd number. **(2 marks)** | D

2 A petrol strimmer costs £92 + $17\frac{1}{2}$% VAT. Work out the total cost of the strimmer. **(2 marks)** | D | **Funct.**

3 These are two 90-day prices plans offered by an electricity supplier.

Plan A: £0.14 per unit of electricity used

Plan B: £0.12 per unit of electricity used + £12.70 fixed charge

In 90 days Mr Vaughan uses, on average, 1200 units of electricity.

Which plan would it be best for Mr Vaughan to use?

You must show all your working. **(3 marks)** | D | AO3 | **Funct.**

4 Rohan wants to find out how much exercise people do in a week.

He has written this question for his survey.

How much exercise do you do in one week? Please tick one box only.

2 to 4 hours ☐ 4 to 6 hours ☐ 6 to 8 hours ☐

a Write down two criticisms of the question. **(2 marks)** | D | AO2 | **Funct.**

b Re-write the question to make it more suitable. **(2 marks)** | C | AO2 | **Funct.**

5 An estate agent collects information on the average prices of three-bedroom houses at certain distances from a motorway. The table shows his results.

Distance from motorway (km)	2	12	5	3	11	15
Average price (£000s)	164	200	176	172	196	214

a Draw a scatter diagram to show this information on a copy of the coordinate grid below. **(2 marks)** D | **Funct.**

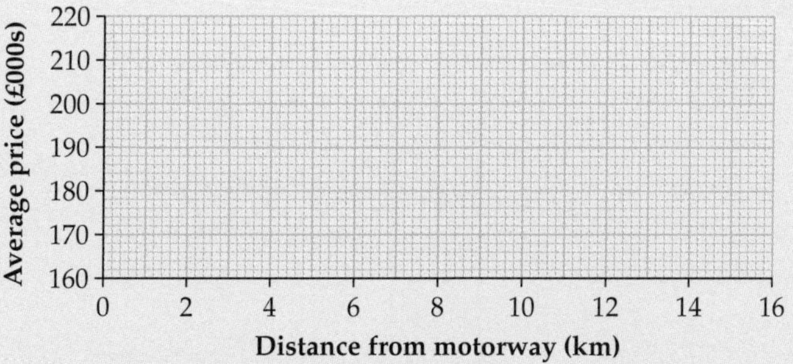

b The estate agent says that the closer a house is to the motorway, the cheaper it is.

Do you agree with this statement? Give a reason for your answer. **(1 mark)** | D | **Funct.**

6* The table shows the number of take-away meals eaten each week by a sample of 60 adults.

Number of take-away meals	0	1	2	3	4	5	more than 5
Frequency	16	12	9	8	5	6	4

a Is it possible to calculate the mean of this data? Give a reason for your answer. **(2 marks)** | D | AO2

b Is it possible to calculate the median of this data? Give a reason for your answer. **(2 marks)** | D | AO2

7 In 2009 the number of people living in a village was 850.

In 2010 the number of people living in the village was 918.

A local councillor says

> 'If the number of people continues to increase by the same percentage each year, by 2015 there will be more than 1200 people in the village.'

Do you agree with the local councillor?

You must show working to support your answer. **(3 marks)** | C | AO3 | **Funct.**

8* Nikki and Sara are business partners. Each year they share a £12 000 bonus in the ratio of the number of years they have been in the business.

This year Nikki has been in the business 6 years and Sara has been in the business 2 years.

Show that in four years' time the difference between the amounts they receive will be halved. **(6 marks)** | C | AO3

9 The speed of light is approximately 300 million metres per second.

a Write this number in standard form. **(1 mark)** | B

b Multiply your answer in part **a** by 60.
Give your answer in standard form. **(1 mark)** | B

c What do you think your answer to part b represents? **(1 mark)** | B | AO2 | **Funct.**

10 This table shows the time taken by 60 adults to complete a puzzle.

Time, t, minutes	Frequency
$0 < t \leqslant 1$	15
$1 < t \leqslant 2$	12
$2 < t \leqslant 3$	23
$3 < t \leqslant 4$	6
$4 < t \leqslant 5$	4

a Draw a cumulative frequency diagram to illustrate this information. **(3 marks)** | B | AO2

b Use your graph to estimate the median. **(1 mark)** | B

c **i** Explain why your answer to part **b** is an estimate. **(1 mark)** | B | AO2

ii Explain what your answer to part **b** represents. **(1 mark)** | B | AO2

d Sue says 'Over 65% of the adults took longer than $2\frac{1}{2}$ minutes to solve the puzzle.'

Is Sue correct? Explain your answer. **(2 marks)** | B | AO2

11 A survey was carried out at a railway station one week to find out how many trains were late, and by how many minutes (to the nearest minute). This histogram shows the results of the survey.

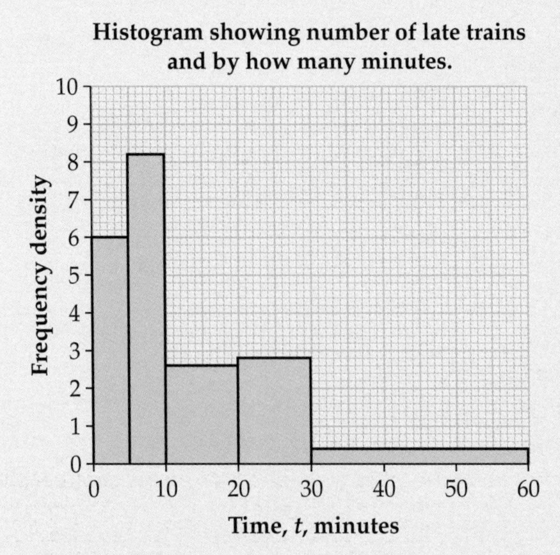

a What is the total number of trains that were late? **(3 marks)** | A

b How many trains were more than 20 minutes late? **(1 mark)** | A

c Work out an estimate of the mean number of minutes a train is late. **(3 marks)** | A* | AO2

12 The table shows the ages of the members of a tennis club.

Age (years)	Under 15	16–30	31–45	Over 45
Number of members	42	28	16	24

The manager of the tennis club wants a stratified sample of 25 people.
How many members should be chosen from each age group? **(3 marks)** | A | **Funct.**

13 A science test is in two parts, a written test and a practical test. Out of all the people who sit the written test, 85% pass. When a person passes the written test, the probability that they pass the practical test is 70%. When a person fails the written test, the probability that they fail the practical test is 65%.
What is the probability that a person chosen at random

a fails both tests **(2 marks)** | A*

b passes exactly one test? **(3 marks)** | A*

Unit 2 Higher Number and Algebra — Non-calculator

1 The formula to find the area, A, of a rectangle is
$A = l \times w$ where l is the length and w is the width of the rectangle.
A rectangle has a length of $2x$ cm and a width of $3x - 1$ cm.

a Show that the formula for the area, A, of the rectangle is
$A = 6x^2 - 2x$ **(1 mark)** | D

b Work out the value of A when $x = 6$. **(2 marks)** | D

2 A two-stage operation is shown.

Input → Subtract 5 → Multiply by 2 → **Output**

a Work out the output when the input is -3. **(1 mark)** | D

b When the input is n what is the output? **(2 marks)** | D

3 The nth term of a sequence is $4n + 5$.
Show that all the terms in the sequence are odd. **(2 marks)** | D | AO2

4* In the first quarter of the year Melia used 1000 units of electricity.
Each unit of electricity cost 10p
In the second quarter of the year Melia used 10% less units, but each unit cost 10% more.
Will Melia's electricity bill be the same in the second quarter of the year as in the first quarter?
You **must** show your working. **(4 marks)** | D | AO2 | **Funct.**

5 An electricity supplier uses this formula to work out the total cost of the electricity a customer uses.
$C = 0.1U + 12$ C is the total cost of the electricity in pounds
U is the number of units of electricity used

a Sam uses 850 units of electricity.
What is the total cost of the electricity she uses? **(2 marks)** | D | **Funct.**

b The total cost of the electricity Clive used is £84
How many units of electricity did he use? **(2 marks)** | C | AO2 | **Funct.**

6 Use approximations to estimate the value of

$$\frac{8105}{21.76 \times 0.219}$$

(3 marks) | C

7 The surface area of a cube is $(6x + 18)\text{ cm}^2$.
Three of these cubes are used to make the cuboid shown.
The surface area of the cuboid is 98 cm^2.
Work out the value of x. **(5 marks)** | C | AO3

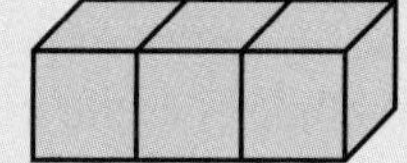

8* Steffan works for a company as a sales representative.
At the start of 2009 he bought a car for £20 000.
At the end of 2009 he sold the car for £15 200.
His mileage for the year was 40 000 miles.
The average cost of running his car was 32p per mile.
His company pay him 40p per mile travelled.
What is Steffan's percentage profit or loss after buying, driving and selling the car?
You **must** show your working and clearly state whether Steffan has made a profit or a loss.
(6 marks) | C | **AO3** | **Funct.**

9 At a school 60% of the students are girls and 40% are boys.
On one particular day, 10% of the girls and 20% of the boys have the flu.
What percentage of the pupils in the whole school have the flu on this particular day? **(3 marks)** | B | **AO2**

10 a Solve the equation $4x + 5 = 23 - 2x$ **(3 marks)** | D

b Solve the equation $\frac{x + 1}{3} + \frac{3x - 5}{2} = 7$ **(4 marks)** | B

11 A is the point (3, 5) and B is the point (1, − 1).
Find the equation of the straight line parallel to AB that passes through the point (4, 2).
You must show your working. **(3 marks)** | B | **AO3**

12 Evan completes a questionnaire.
He ticks this box to show the values his age lies between.

✓ $15 \leq a < 20$

a Write down all the whole number ages that Evan could be. **(1 mark)** | D

b Show the inequality $15 \leq a < 20$ on a copy of this number line.

12 13 14 15 16 17 18 19 20 21 22

(1 mark) | D

Evan has two brothers, Alun and Berwyn.
The sum of Alun and Berwyn's ages is 44.
The difference is 6.

c Write equations for the sum of their ages and the difference in their ages. **(1 mark)** | B

d Solve the equations simultaneously to find the ages of Alun and Berwyn. **(2 marks)** | B

13 Simplify fully $\frac{4x^2 - 25}{2x^2 - x - 15}$ **(5 marks)** | A

14 a Simplify fully $\frac{(x^3)^4}{x^2}$ **(2 marks)** | A

b Explain why $125^{-\frac{1}{3}} = \frac{1}{5}$ **(2 marks)** | A* | **AO2**

15 a Write the expression $x^2 - 4x - 10$ in the form $(x + a)^2 + b$. **(3 marks)** | A*

b Hence, or otherwise, solve $x^2 - 4x - 10 = 0$.
Give your answers in surd form. **(2 marks)** | A*

16 Write the expression $(3 + \sqrt{8})(7 - \sqrt{18})$ in the form $a + b\sqrt{c}$, where a, b and c are integers. **(4 marks)** | A*

Unit 3 Foundation Geometry and Algebra — Calculator allowed

1 a Shona is facing south. She makes a $\frac{1}{4}$ turn clockwise.
In which direction is she now facing? **(1 mark)** | G

b How many degrees are there in a $\frac{1}{4}$ turn clockwise? **(1 mark)** | G

c This diagram shows a seven sided shape.

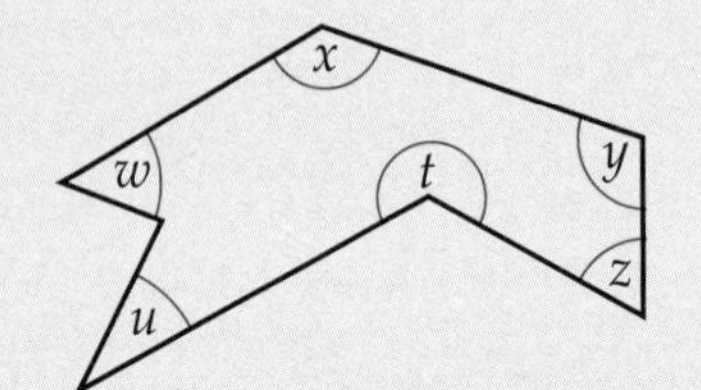

i How many acute angles does the shape have? **(1 mark)** | G

ii What type of angle is angle x? **(1 mark)** | G

iii What type of angle is angle t? **(1 mark)** | F

2 The diagram shows a circle.

Fill in the missing words in these sentences.

a The line *AB* is called the .... **(1 mark)** | G

b The curved line *BC* is called an .... **(1 mark)** | G

c The shaded area is called a .... **(1 mark)** | G

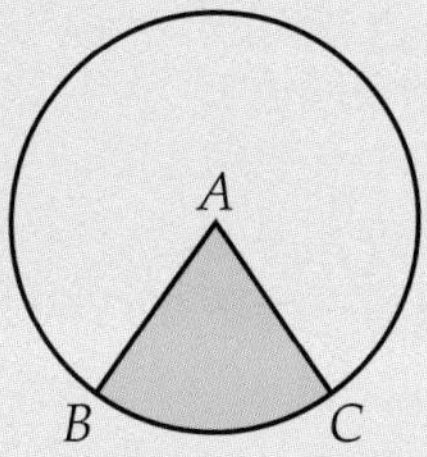

3* Ted has a maximum of £400 to spend on a holiday.

He needs to pay for a hotel, flights and insurance.

He sees this advert.

Can Ted afford this holiday?

You **must** show your working.

Holiday to Spain - Special Offer!
Hotel: £287 per person
Flights: £125 per person
Insurance: £19 per person
Buy online and get 10% off total price.

(2 marks) | G | **AO1 (2 marks)** | F | **AO2 Funct.**

4 a What name is given to this type of triangle?

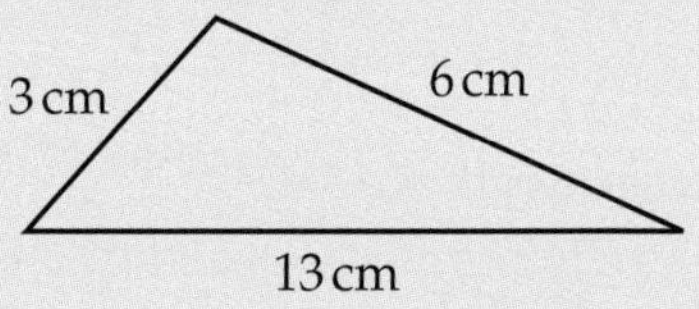

3 cm, 6 cm, 13 cm

(1 mark) | G

b Copy the following triangles and draw on all their lines of symmetry.

i

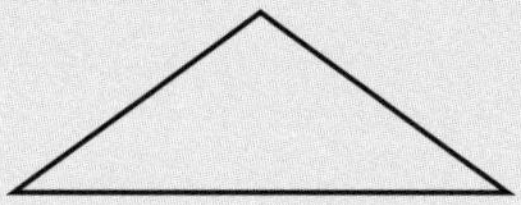

(1 mark) | G

ii

(2 marks) | G

c Draw a rectangle 5 cm long and 2 cm wide. By drawing three extra lines, show how you can divide the rectangle into four right-angled triangles. **(2 marks)** | G

5 Write down the amount shown on each of these scales.

a

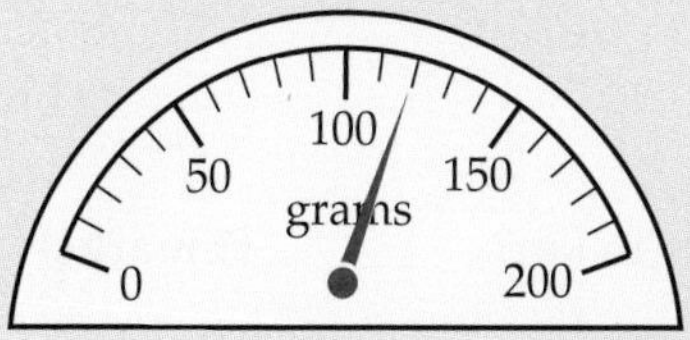

0 50 100 150 200 grams

(1 mark) | G | **Funct.**

b

°C −20 −10 0 10 20 30 40

(1 mark) | G | **Funct.**

c Write down the two different times that clock could be showing using the 24-hour clock.

(2 marks) | G | **Funct.**

6 Which two of the following could be the net of a cube?

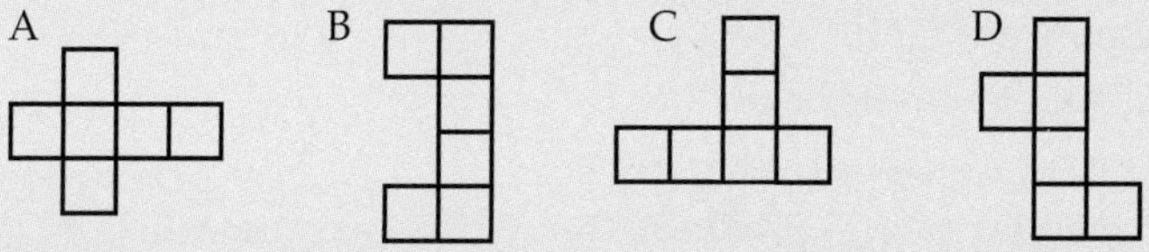

A B C D

(2 marks) | G

7 Sascha is going buy pet insurance for his dog.
There are two ways of paying.

Pet Insurance	
Monthly payment	£4.55
Yearly payment	£47.50

How much would Sascha save by making a single yearly payment? **(3 marks)** | G | **AO2** | **Funct.**

8 a Here is a quadrilateral.

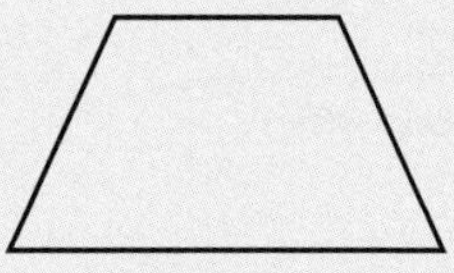

i Write down the name of this quadrilateral. **(1 mark)** | F

ii Write down the order of rotational symmetry of this quadrilateral. **(1 mark)** | F

b Here is a different quadrilateral.

i Write down the name of this quadrilateral. **(1 mark)** | F

ii Write down the order of rotational symmetry of this quadrilateral. **(1 mark)** | F

9 Work out the size of angle x in this diagram.

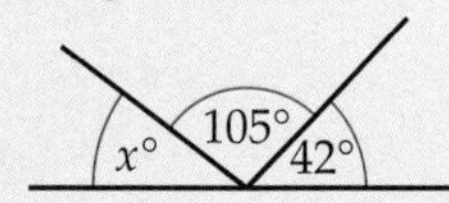

(2 marks) | F

10* This table shows some distances in miles and their equivalent distances in kilometres.

Miles	0	10	20	30	40
Kilometres	0	16	32	48	64

a Draw a graph to show this information.
Plot miles on the x-axis going from 0 to 40, and kilometres on the y-axis going from 0 to 70. **(3 marks)** | F

b Simon drives 36 km.
Use your graph to work out how many miles is this? **(1 mark)** | F

c The distance from Leeds to Chester is 70 miles.
Use your graph to work out the distance from Leeds to Chester in km? **(3 marks)** | F | **AO2**

11* Sam plots the points A, B and C on the grid shown.

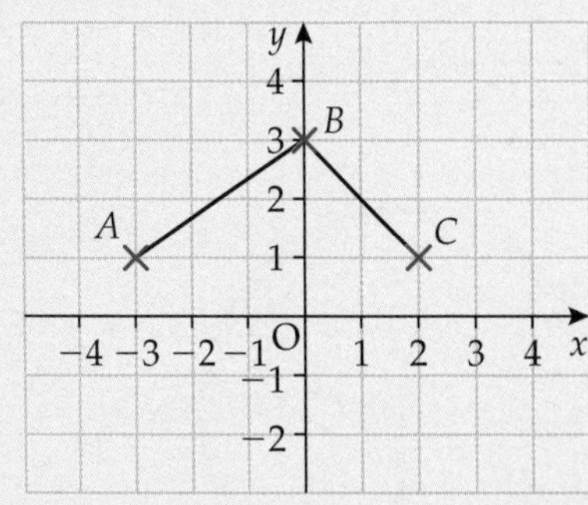

a Sam says 'If I plot point D at $(-1, -1)$ shape $ABCD$ will be a kite'.
Explain why Sam is wrong. **(1 mark)** | E | **AO2**

b Write down the coordinates of the point D, that would make shape $ABCD$ a kite. **(1 mark)** | E | **AO2**

12* A bag of sweets contains 60 sweets to one significant figure.
What is the smallest number of sweets that could be in the bag?
Circle your answer from the list below and give a reason for your answer.

59 sweets   54 sweets   55 sweets   57 sweets **(2 marks)** | E | **AO2**

13 The diagram shows the position of two buoys, A and B.

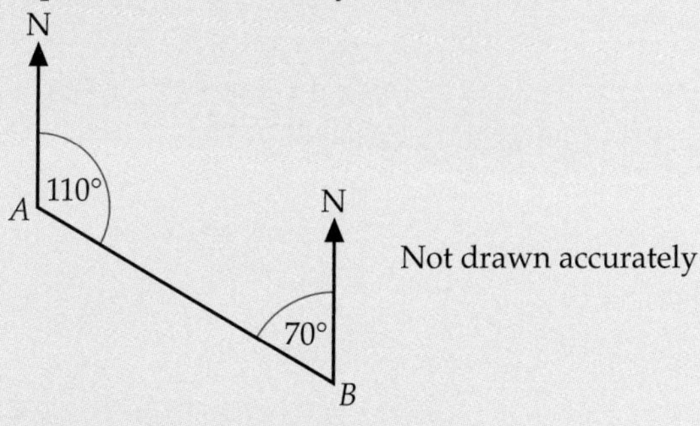

Not drawn accurately

a Write down the bearing of B from A. **(1 mark)** | E

b Work out the bearing of A from B. **(2 marks)** | D

14 Franz flies from the UK to Germany then from Germany to the USA.

Before he leaves the UK he changes £500 into Euros.

The exchange rate is £1 = €1.08 (euros).

In Germany he spends €135.

When he arrives in the USA he changes his remaining euros into dollars ($).

The exchange rate is $1 = €0.72.

How many dollars does he get? **(4 marks)** | E | **AO2** | **Funct.**

15 In this quadrilateral the largest angle is 128°.

The opposite angle is 32° less than the largest angle.

One of the other angles is half the size of the largest angle.

Work out the size of the remaining angle. **(3 marks)** | E | **AO3**

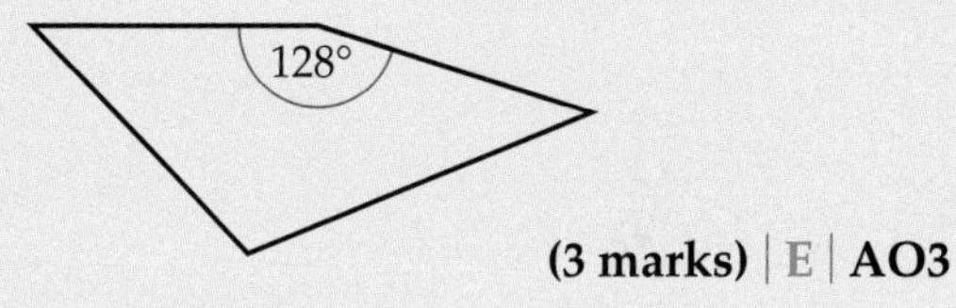

16 The diagram shows a path made from paving slabs.

The paving slabs are identical.

The length of a paving slab is twice as long as the width.

The total length of the path is 5.6 m.

What is the total area of the path? **(4 marks)** | E | **AO3**

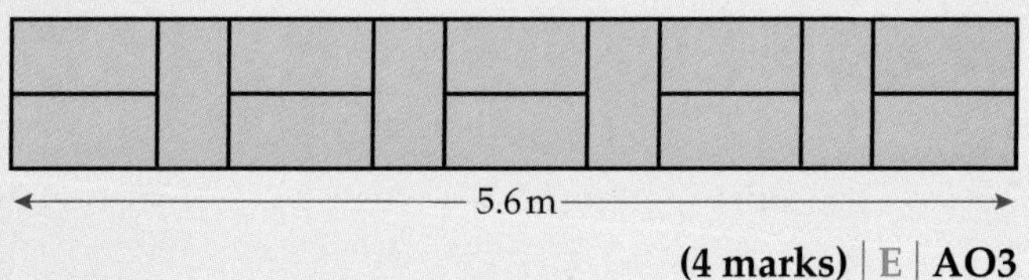

17 Copy this diagram. Draw the image of triangle A after a reflection in the line $x = -1$.

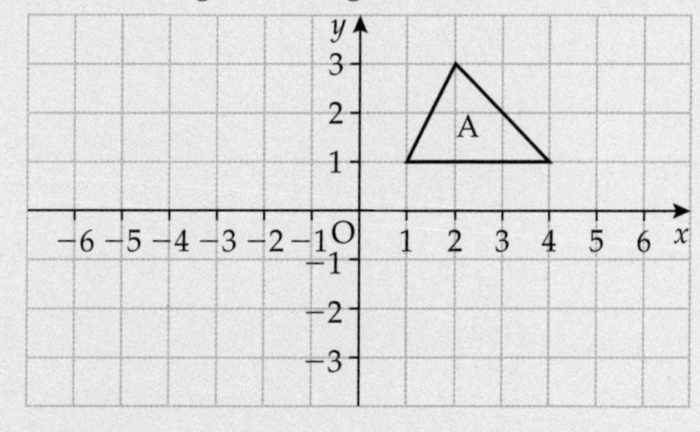

(2 marks) | D

18 The diagram shows a pentagon.

Work out the size of angle x.

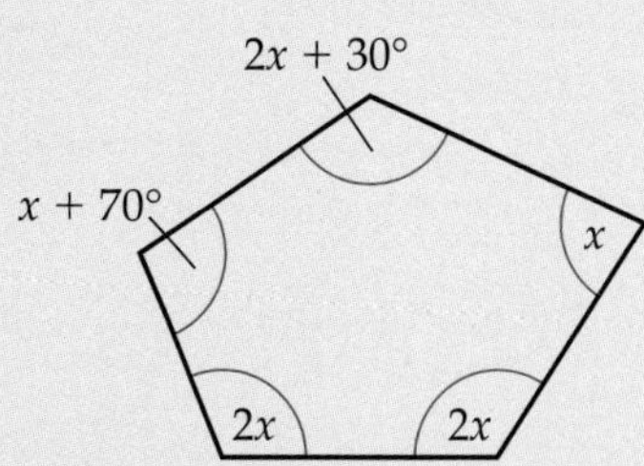

(4 marks) | D | **AO2**

19 To go on one of the rides at a theme park you have to be over 1.4 m tall.

Alison is 4 feet 5 inches tall. There are 12 inches in 1 foot.

Can Alison go on the ride?

You **must** show all your working. **(4 marks)** | D | **AO3**

20 Calculate the area of a semicircle with diameter 12 cm.
Leave your answer in terms of π. **(2 marks)** | C

21 a On a coordinate grid, draw the graph of $y = x^2 - 2x - 3$ for $-2 \leqslant x \leqslant 4$.
Draw the x-axis going from -2 to $+4$ and the y-axis going from -6 to $+6$. **(3 marks)** C

b Use your graph to solve the equation $x^2 - 2x - 3 = 2$ **(2 marks)** | C

22 This triangular prism has a volume of 918 cm^3.
Work out the length, x, of the prism.
You **must** show your working.

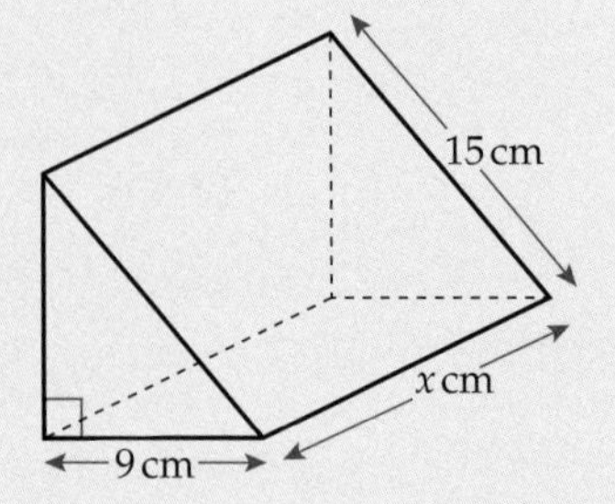

(5 marks) | C | AO3

Linear Paper 1 Higher tier

Non-calculator

1* **a** There are 50 members in a scuba diving club.
22 of the members are women.
What percentage of the members are men? **(3 marks)** | D

b The rate of VAT was raised in January 2010 from 15% to $17\frac{1}{2}$%.

Special offer!
Cement mixer £380 + VAT

Work out the difference in price of a cement mixer due to the increase in VAT.
(3 marks) | D | AO2 | **Funct.**

2 Enid is travelling to Cumbria for her holiday.
She begins her journey at 9 am and travels 80 km in the first hour.
Between 10 am and 11 am she travels only 50 km due to heavy traffic.
At 11 am she stops for a half-hour break.
She reaches her destination after a further $2\frac{1}{2}$ hours and a distance of 150 km.

a Draw a distance–time graph of Enid's journey. **(4 marks)** | D | AO2

b During which section of her journey was Enid travelling the fastest?
Explain how you know. **(1 mark)** | D

c Work out Enid's average speed for the whole journey. **(3 marks)** | D | **Funct.**

3 A shoe shop records the number of each size shoe it sells.
These are the results for one week.

Shoe size	Frequency
4	1
5	14
6	17
7	0
8	8

a Write down the range of this data. **(1 mark)** | D

b Write down the mode of this data. **(1 mark)** | D

c Calculate the mean shoe size sold. **(3 marks)** | D

4 Here is a trapezium.

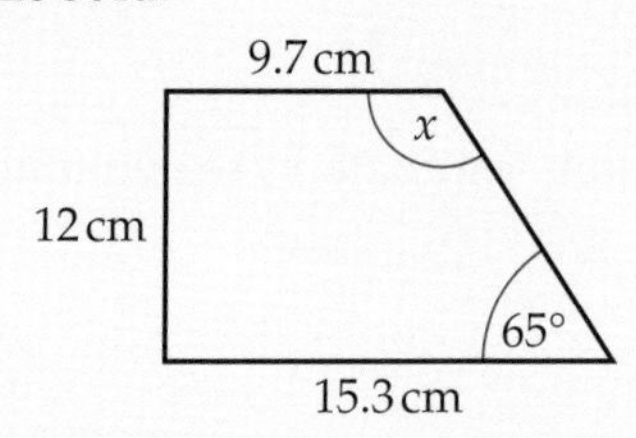

a State the value of x. **(1 mark)** | D

b Work out the area of the trapezium. **(2 marks)** | D

5* John is a builder. He mixes his own concrete out of sand and cement.
The table shows the sand : cement ratios for different types of concrete.

Type of concrete	Sand : cement
general building (above ground)	5 : 1
general building (below ground)	3 : 1
internal walls	8 : 1

John is starting a new job. He estimates that he needs 240 kg of concrete for general building above ground and 180 kg of concrete for internal walls.
Sand and cement are both sold in 25 kg bags.
Work out how many bags of sand and how many bags of cement John needs to buy.
(6 marks) | C | AO2 | **Funct.**

6 The table shows the distances, d km, that some people cycle to work.

Distance, d (km)	Frequency
$0 \leqslant d < 5$	13
$5 \leqslant d < 10$	9
$10 \leqslant d < 15$	5
$15 \leqslant d < 20$	2
$20 \leqslant d < 25$	3

a Which class interval contains the median?
Explain how you worked out your answer. **(2 marks)** | C

b Explain why it is not possible to calculate the exact mean distance. **(1 mark)** | C | **AO2**

7 Dylan starts with this shape. He transforms the shape to make this pattern.

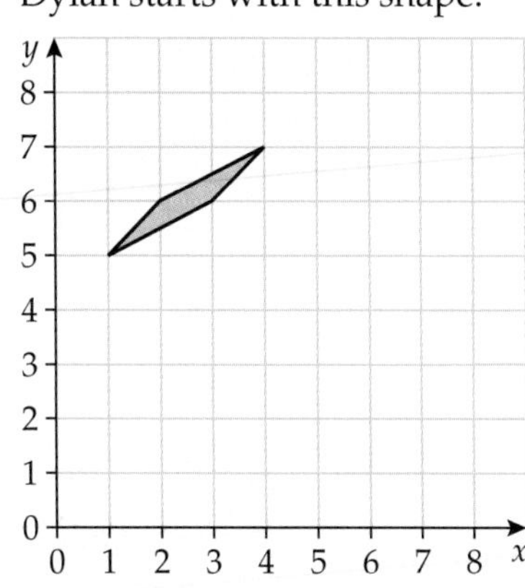

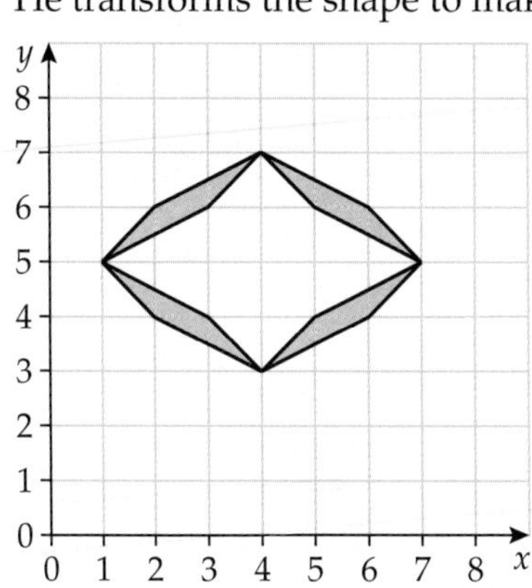

Describe the transformations he uses. **(4 marks)** | C | **AO3**

8* The diagram shows a solid. The lengths x, y and z are shown.

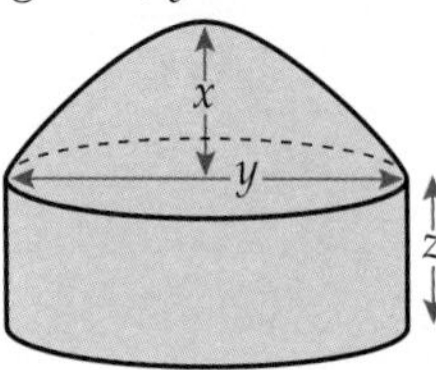

One of the following formulae may be used to estimate V, the volume of the solid.

$V = 4x + 2y + 3z$

$V = 4x^2y + 2x^2z$

$V = 2x(3y^2 + 4z)$

a Explain why the formula $V = 4x + 2y + 3z$ cannot be used to estimate the volume of the solid. **(1 mark)** | B

b State, with a reason, which of the above formulae may be used to estimate the volume of the solid. **(2 marks)** | B

9* **a** Factorise $x^2 - 49$ **(2 marks)** | B

b **i** Show that $(x + y)^2 - (x - y)^2 \equiv 4xy$ **(2 marks)** | B | **AO2**

ii Use the identity in part **i** to work out $23^2 - 17^2$ **(2 marks)** | B | **AO2**

c Prove that the sum of three consecutive integers is a multiple of 3. **(3 marks)** | B | **AO3**

10 A report into use of plastic bags in the UK states that approximately 7 800 million plastic bags are used in the UK each year.

a Write this number in standard form. **(1 mark)** | B

b The population of the UK is approximately 6×10^7.
What is the mean number of plastic bags used per person in the UK? **(2 marks)** | B | **Funct.**

11 AC is a tangent to the circle at B.
Angle $EOD = 210°$
Angle $EDB = 48°$.

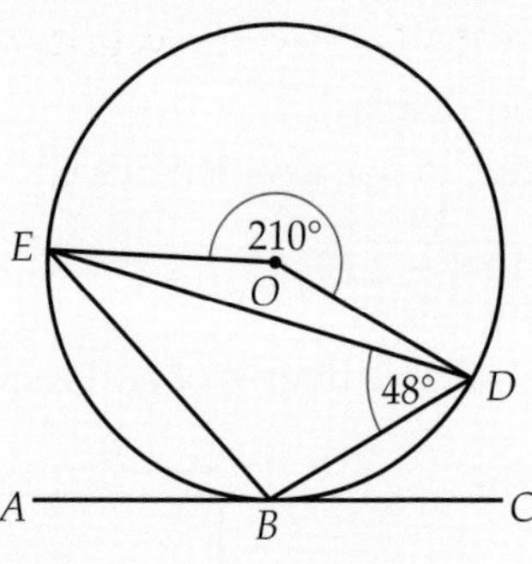

a Give a reason why angle $EBD = 105°$. **(1 mark)** | A

b Work out the value of angle DBC. **(2 marks)** | A

12 The diagram shows a major segment of a circle of radius 15 cm.
The length of the major arc is 27π cm.

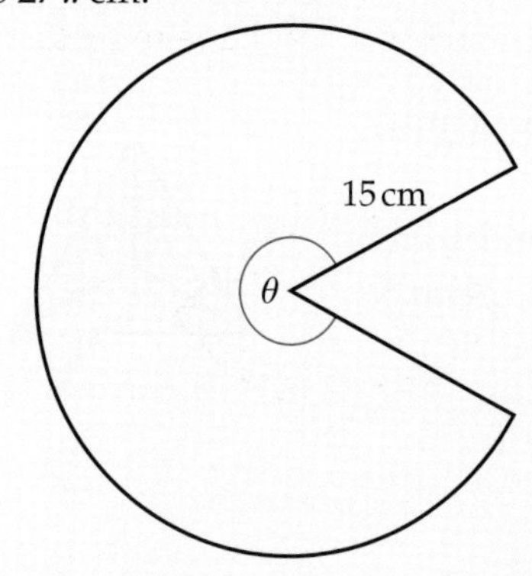

Calculate the angle of the major segment, marked θ on the diagram. **(3 marks)** | A

13 A is the point (3, −2) and B is the point (6, 10).
Work out the equation of the line parallel to AB that passes through the point (1, 9). **(4 marks)** | A | AO3

14 Simplify $64^{-\frac{2}{3}}$ **(3 marks)** | A*

15 a Write the expression $x^2 - 6x - 23$ in the form $(x + p)^2 + q$. **(2 marks)** | A*

b Hence solve the equation $x^2 - 6x - 23 = 0$.
Give your answer in the form $a \pm b\sqrt{2}$. **(3 marks)** | A*

16 A bag contains three white and four red discs.
Anil takes a disc at random from the bag. He does not replace it.
Bern then takes a disc at random from the bag.
Anil wins if both the discs are the same colour.
Who has a better chance of winning?
You must show working to support your answer. **(3 marks)** | A* | AO3

Linear Paper 2 Higher tier

Calculator allowed

1 Andy and Bari share £240 in the ratio 1 : 3.
How much does Bari get? **(3 marks)** | D

2 Steven goes shopping. He has £80 to spend.
He spends £24.50 on a pair of squash shoes and £6.99 on some squash balls.
He sees this advert for a squash racket.

Squash Racket: normal price £95
Special offer!
45% off normal price

Does Steven have enough money to buy the squash racket?
You **must** show your working. **(4 marks)** | D | AO2 | Funct.

3 **a** Use your calculator to work out $\sqrt[3]{117}$.

i Write down your full calculator display. **(1 mark)** | D

ii Write down your answer correct to one decimal place. **(1 mark)** | D

b Use your calculator to work out $\dfrac{9.49}{1.88 + 2.27}$

i Write down your full calculator display. **(1 mark)** | D

ii Write down your answer to a suitable degree of accuracy. **(1 mark)** | D

4 This is the area of a garden that is going to have turf laid to make a new lawn.

Turf costs £2.25 per square metre.

It can only be bought in whole numbers of square metres.

Work out the cost of the turf for the new lawn.

You **must** show your working.

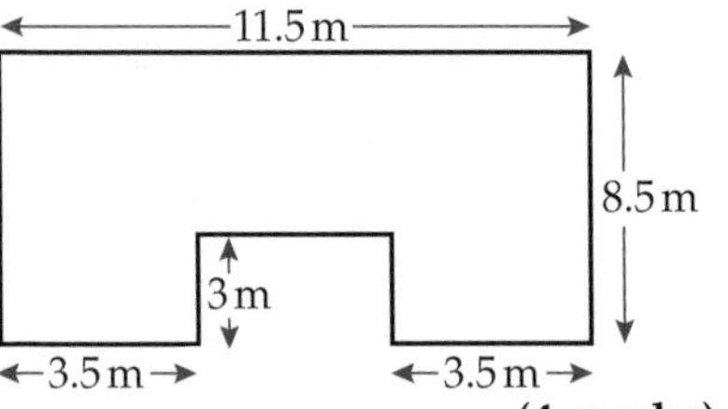

(4 marks) | D | AO2 | **Funct.**

5* Megan designs a game to raise money for charity.

She makes two fair spinners.

One is four-sided and numbered 1, 2, 3 and 4.

The other is three-sided and numbered 1, 2 and 3.

Players spin the spinners at the same time and multiply the numbers together.

Each player pays £1 to play the game.

If a player gets a score of 8 they win £3.

If a player gets a score less than 3 they win £2.

If 300 people play the game, how much money should Megan expect to make?

Show clearly how you worked out your answer. **(6 marks)** | D | AO3

6 **a** Factorise $6y + 18$ **(1 mark)** | D

b Solve $\frac{x}{7} = -8$ **(1 mark)** | D

c $d \triangle e$ means $9d + 7e$

When $x \triangle 4 = 6 \triangle x$, work out the value of x. **(3 marks)** | D | AO2

d Expand and simplify $(x - 7)(x + 3)$ **(2 marks)** | C

7 Maria runs a business making tin containers.

She uses this formula to work out the approximate total surface area of a cylindrical tin.

$A = 6r(r + h)$	where: A is the total surface area in cm^2 r is the radius of the tin in cm h is the height of the tin in cm

a Use the formula to work out the approximate total surface area of a cylindrical tin with radius 3 cm and height 9 cm. **(2 marks)** | D | **Funct.**

b Rearrange the formula to make h the subject. **(2 marks)** | B | **Funct.**

8 $ABCD$ is a quadrilateral.

Angle D is 90°

Angle $A = x$, angle $B = 3x - 20°$ and angle $C = 2x + 32°$

Work out the largest angle in the quadrilateral.

You **must** show your working.

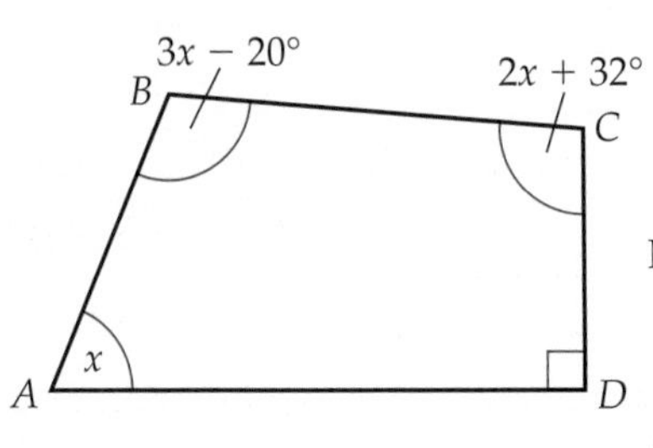

Not drawn accurately

(5 marks) | D | AO3

9 Harry is going to carry out a survey on visits to the dentist.
These are two of the questions he has written.

1. Do you agree that it takes to long to get am appointment to see the dentist?

Strongly agree ☐ Agree ☐ Don't know ☐

2. How often do you visit the dentist?

a Give two reasons why question 1 is unsuitable. **(2 marks)** | C

b Give a suitable response section for question 2. **(1 mark)** | C

c Harry decides to carry out his survey outside his local dentist surgery.
Explain why this sample is likely to be unrepresentative. **(1 mark)** | C | **AO2**

10 a Complete the table of values for $y = x^2 - 2x - 4$. **(2 marks)** | C

x	−3	−2	−1	0	1	2	3	4
y		4	−1	−4		−4	−1	

b Draw a coordinate grid that goes from −3 to +4 on the x-axis and −5 to +15 on the y-axis.
On the grid, draw the graph of $y = x^2 - 2x - 4$ for values of x from −3 to +4. **(2 marks)** | C

c Use your graph to write down the solutions to the equation $x^2 - 2x - 4 = 0$ **(2 marks)** | C

11 M is the point (5, 6) and N is the point (9, 14).

a Work out the midpoint of the line MN. **(2 marks)** | C

b Calculate the length of the line MN.
Give your answer correct to one decimal place. **(2 marks)** | C | **AO2**

12* a Show that $4(4y - 7) - 5(2y - 9) \equiv 3(2y + 7) - 4$ **(4 marks)** | C | **AO3**

b Solve the equation $\dfrac{3x+4}{2} - \dfrac{2x+1}{3} = 5$ **(4 marks)** | B

13* A company produce two types of light bulb, type A and type B.
The lifetime, in hours, of a sample of 80 of each type of bulb was measured.
The cumulative frequency diagram shows the results for the type A light bulb.

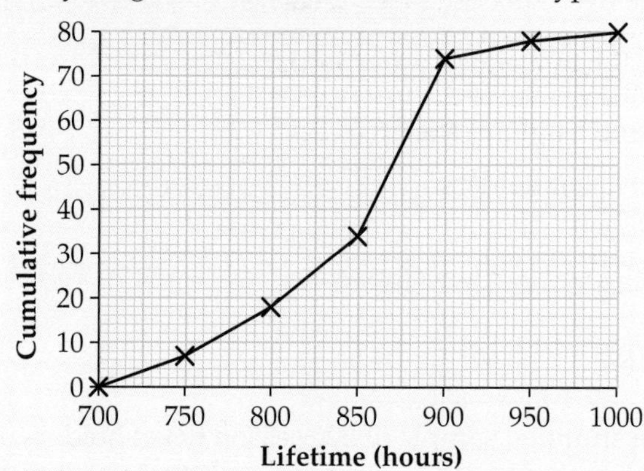

a Estimate the median and interquartile range for the type A light bulb. **(3 marks)** | B | **Funct.**

b The median and interquartile range for the type B light bulb are 825 hours and 110 hours respectively.
Which type of bulb is longer lasting?
Use the medians and inter-quartile ranges to clearly explain your answer.
(2 marks) | B | **AO2** | **Funct.**

14 A catering company cooks meals for parties.

They offer three main courses: lasagne (L), fish (F) or quiche (Q).

To accompany the main course they offer either salad (S) or chips (C).

The company use previous data to estimate the number of different types of meals they need to cook. The probability of a person choosing lasagne is 0.4 and fish is 0.5.

The probability of a person choosing salad is 0.25.

a Copy and complete the tree diagram to show all the possible outcomes.

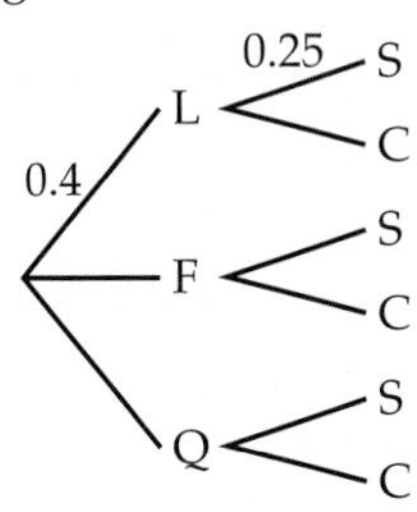

(2 marks) | B | AO2 | **Funct.**

b Work out the probability that a person chooses fish and chips. **(2 marks)** | B | **Funct.**

c At the next party, 200 guests are expected.

Estimate the number of quiche and salad meals the company will need to cook.

(2 marks) | B | AO2 | **Funct.**

15 The diagram shows some angles around a point.

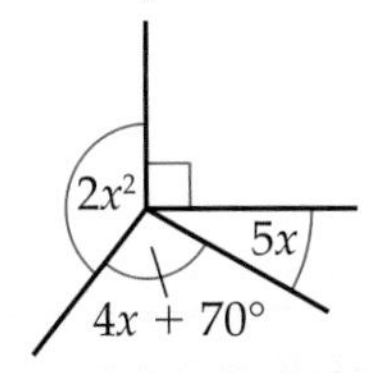

a Show that x satisfies the equation $2x^2 + 9x - 200 = 0$ **(2 marks)** | B | AO2

b Solve the equation $2x^2 + 9x - 200 = 0$.

Hence work out the sizes of the angles around the point. **(4 marks)** | A

16 Serge wants to replace the old tiles on his garage roof with new clay tiles.

The recommended minimum angle of elevation of a roof suitable for clay tiles is 35°.

The diagram shows the dimensions of his roof.

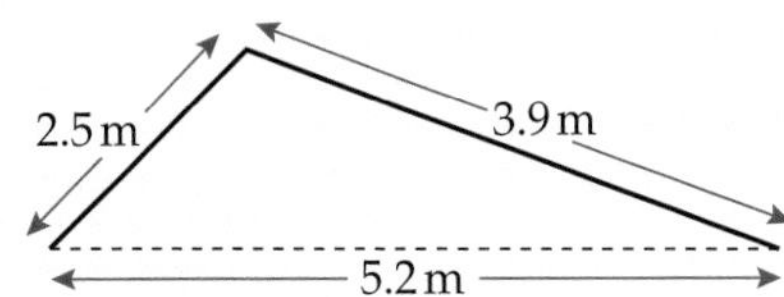

Should Serge use clay tiles on his garage roof?

You **must** show working to support your answer. **(3 marks)** | A | AO2 | **Funct.**

17 A football club has 28 000 season ticket holders.

They are classified by age as follows.

Age (years)	Under 21	21–40	41–60	Over 60
Number of season ticket holders	6440	12 680	7250	1630

The football club wants to take a stratified sample of 500 season ticket holders.

Calculate the number that should be sampled from each age group. **(3 marks)** | A | **Funct.**

18 Two cones are mathematically similar.

The smaller cone has a volume of 120 cm³.

The larger cone has a volume of 405 cm³.

The curved surface area of the smaller cone is 102 cm².

Work out the curved surface area of the smaller cone. **(4 marks)** | A | AO3

19 Will draws a circle at two of the opposite vertices of a rhombus.

The rhombus has a side length of 6 cm.

The diagram shows the shape Will has drawn.

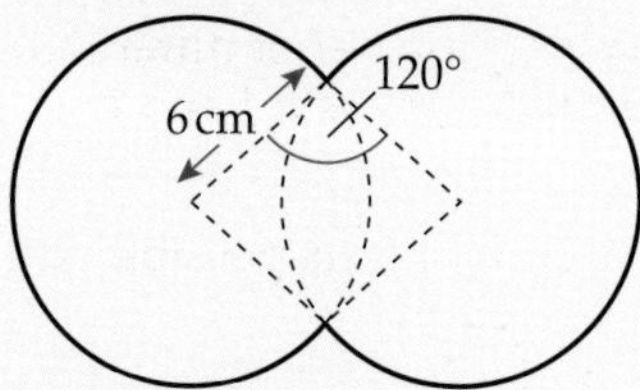

a Show that the perimeter of the shape is 20π cm. **(3 marks)** | A | AO2

b Work out the area of the shape. **(5 marks)** | A* | AO3

20 OAB is a triangle with $\overrightarrow{OA} = \mathbf{a}$ and $\overrightarrow{OB} = \mathbf{b}$.

M is the mid-point of OB and P is the point on AB such that $AP : PB = 1 : 2$.

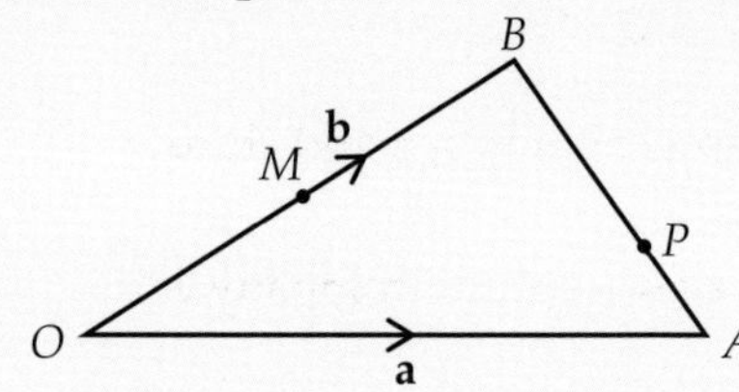

Find expressions for these vectors, in terms of **a** and **b**, simplifying your answers when possible.

a $\overrightarrow{AB}$ **(1 mark)** | A

b $\overrightarrow{OM}$ **(1 mark)** | A

c $\overrightarrow{OP}$ **(2 marks)** | A*

d $\overrightarrow{MP}$ **(2 marks)** | A*

21 Solve these simultaneous equations using an algebraic method.

$$y - 5x = 2$$

$$y = x^2 + x - 10$$

(5 marks) | A*

1